25905
5-11-91
IN.

Knowledge Representation

An Approach to Artificial Intelligence

This is volume 32 in the A.P.I.C. Series

General Editors: M.J.R. Shave *and* I.C. Wand
A complete list of titles in this series appears at the end of this volume

The A.P.I.C. Series
No 32

Knowledge Representation

An Approach to Artificial Intelligence

T.J.M. BENCH-CAPON

*Department of Computer Science,
University of Liverpool, England*

ACADEMIC PRESS

Harcourt Brace Jovanovich, Publishers
London   San Diego   New York   Boston
Sydney   Tokyo   Toronto

ACADEMIC PRESS LIMITED
24/28 Oval Road,
London NW1 7DX

United States Edition published by
ACADEMIC PRESS INC.
San Diego, CA 92101

Copyright © 1990 by
ACADEMIC PRESS LIMITED

This book is printed on acid-free paper

All Rights Reserved

No part of this book may be reproduced in any form by photostat, microfilm, or any other means, without written permission from the publishers

British Library Cataloguing in Publication Data

Is available

ISBN 0-12-086440-1

Q
335
B45
1990

Typeset by Colset, Pte. Ltd., Singapore
Printed in Great Britain by
Galliard (Printers) Ltd, Great Yarmouth, Norfolk

Preface

This book is about knowledge representation. Knowledge representation has emerged as one of the fundamental topics in the area of computer science popularly known as *artificial intelligence* (AI). The reason for this is simple: the basic idea is that intelligence, whatever else it involves, does at least involve knowing things, and exploiting them so as to respond appropriately to a given situation. Therefore it seems reasonable to suppose that if we wish to make an intelligent computer system we must have a way of getting it to know things, and that involves finding a way of representing the things we wish it to know so that they can be encoded within the computer system. Thus the intention of the book is to provide an introduction to the field of artificial intelligence, via an examination of issues concerning knowledge representation.

Whilst a variety of paradigms that have been used in artificial intelligence will be considered, there will be some focus on those which might be termed *rule based* at the expense of those which might be termed *structured object* representations. This emphasis is in part due to my personal predilections, but none the less I have tried to give reasonable coverage also to competing ideas. The first four chapters will be concerned with general issues, and will provide background to the rest of the book, which will comprise a detailed consideration of three of the major knowledge representation paradigms which have come out of AI research, and thus represent much of the cumulative wisdom on the matter.

In the first chapter, we will consider what is meant by "artificial intelligence", and the related terms like *knowledge-based systems* and *expert systems*, so that we can be clear about our motivation, and about exactly what we are trying to achieve with such programs. The second chapter gives a general consideration of what knowledge representation is, and what criteria we can use to assess a particular form of knowledge representation. The third chapter outlines some of the basic concepts and terminology of formal logic, which is essential as a means of describing and distinguishing between the various computer-oriented paradigms discussed later. The fourth chapter gives a description of the key topics relating to *search*, which is historically important in AI, and which is also crucial to the understanding of a range of knowledge representation issues.

Chapters 5, 6 and 7 consider the three leading representation paradigms—production rules, structured object representations and predicate calculus,—so that we can see the style of systems they produce, and consider their various strengths and weaknesses. To illustrate some of the use that has been made of these paradigms the eighth chapter examines PROLOG, which can be seen as a practical instantiation of many of the ideas deriving from the predicate logic paradigm, and the ninth chapter discusses *expert systems*, which represent an area of application which has made extensive use both of production rules and structured object representations.

In the last chapter a number of issues relating to the use and limitations of the paradigms will be discussed, so as to indicate some of the unsolved problems of knowledge representation.

The book originated in a course of lectures given by me to third-year undergraduates in computer science at the University of Liverpool and is directed primarily at undergraduates embarking on a first course in artificial intelligence. Whilst I have tried to presuppose as little knowledge as possible, a reader will need some familiarity with the concepts of computer science, and would be helped by such grounding in formal logic as might be given by an introductory course in the subject.

I would like to take this opportunity to thank all those with whom I have discussed knowledge representation issues: particularly Marek Sergot, Robert Kowalski and my former colleagues at Imperial College, Justin Forder and my former colleagues on the Alvey–DHSS Demonstrator project, and my present colleagues at the University of Liverpool. Thanks are due also to those who have read and commented on earlier drafts, especially my wife, Priscilla, who contributed many valuable comments which improved the grammar, spelling and comprehensibility. Needless to say, remaining errors and unsupported prejudices are my own.

Contents

Preface			v
1. Introduction to Artificial Intelligence and Knowledge-Based Systems			1
1.1.	Artificial and intelligent		1
	1.1.1.	Popular views of intelligent machines	2
	1.1.2.	Human versus machine intelligence	3
	1.1.3.	Turing test	4
	1.1.4.	Typical AI applications	5
	1.1.5.	AI and understanding the human mind	6
	1.1.6.	Techniques used in AI	7
	1.1.7.	Different motivations for AI	7
	1.1.8.	Knowledge-based systems	8
2. Introduction to Knowledge Representation			11
2.1.	What is knowledge representation?		11
	2.1.1.	Purpose of knowledge representation	12
2.2.	Criteria of adequacy		13
	2.2.1.	Metaphysical adequacy	13
	2.2.2.	Epistemic adequacy	14
	2.2.3.	Heuristic adequacy	14
	2.2.4.	Computational tractability	15
2.3.	Expressiveness		15
	2.3.1.	Lack of ambiguity	16
	2.3.2.	Clarity	16
	2.3.3.	Uniformity	16
	2.3.4.	Notational convenience	17
	2.3.5.	Relevance	17
	2.3.6.	Declarativeness	17
2.4.	Example of the use of these criteria		18
2.5.	Major paradigms		18
2.6.	Manipulation of representations		19
	2.6.1.	Deduction	21
	2.6.2.	Induction	22
	2.6.3.	Abduction	23
	2.6.4.	Other methods of reasoning	24

Contents

3. Logic — 27
   3.1. Basics — 27
   3.2. Basics of propositional calculus — 28
      3.2.1. Notation of propositional calculus — 29
      3.2.2. Truth tables — 30
      3.2.3. Assignments and models — 31
      3.2.4. Nature of a proof — 32
      3.2.5. Natural deduction — 32
      3.2.6. Equivalences between the logical connectives — 33
      3.2.7. Monotonicity — 33
   3.3. Basics of predicate calculus — 34
      3.3.1. Predicates — 34
      3.3.2. Notation — 34
      3.3.3. Propositional functions and quantifiers — 35
      3.3.4. Quantifiers and natural deduction — 36
      3.3.5. Relations — 36
      3.3.6. Identity — 37
      3.3.7. Putting expressions into normal forms — 38
      3.3.8. Method of transformation — 38
   3.4. Exotic logics — 39
   3.5. Modal logics — 40

4. Search — 41
   4.1. Search spaces — 41
      4.1.1. Characterisation of search problem — 42
      4.1.2. Examples — 42
   4.2. Search methods — 44
      4.2.1. Goal-driven and data-driven search — 44
      4.2.2. Breadth-first, depth-first, and bounded depth-first searches — 46
      4.2.3. Heuristic or ordered search — 51
      4.2.4. The 8-puzzle as an example of heuristic search — 52
   4.3. Limitations of search — 55
      4.3.1. Search space size — 55
      4.3.2. Evaluation function problems — 55
   4.4. Human problem solving — 57
      4.4.1. Decomposition of problems — 57
      4.4.2. Search for patterns — 58
      4.4.3. Rules of thumb — 59

5. Production Rules — 63
   5.1. Form of production rules — 63
      5.1.1. Entity–attribute–value triples — 63
      5.1.2. Variables in production rules — 64
   5.2. Components of a production system — 66
      5.2.1. Working memory — 66
      5.2.2. Production memory — 67
      5.2.3. Rule interpreter — 68
   5.3. Operation of a production system — 68
      5.3.1. Goal-driven search using production rules — 69
      5.3.2. Data-driven use of production rules — 70

			Contents	ix

		5.3.3.	An example	70
		5.3.4.	Control and conflict resolution	71
		5.3.5.	Conflict resolution strategies	72
		5.3.6.	Defeasible rules	74
	5.4.	Pros and cons of production systems		75

6. Structured Objects 79
   6.1. Semantic networks 79
      6.1.1. Origins 80
      6.1.2. Features 81
      6.1.3. Inference in semantic nets 83
      6.1.4. Psychological justification 85
   6.2. Frames 86
      6.2.1. Basic ideas 87
      6.2.2. Use of frames 89
      6.2.3. Inheritance hierarchies 90
      6.2.4. Default values 93
      6.2.5. Multiple inheritance 94
      6.2.6. Object-orientated knowledge representation 94
      6.2.7. Frames combined with rules 98
      6.2.8. Combining frames with logic—KRYPTON 99
      6.2.9. AI toolkits 100

7. Logic and Predicate Calculus 103
   7.1. Advantages of predicate calculus 104
   7.2. Foundations of logic programming 105
      7.2.1. Resolution 105
      7.2.2. Control of general resolution 108
      7.2.3. Horn Clauses 110
      7.2.4. Resolution and predicate calculus 111
      7.2.5. Skolemisation 112
      7.2.6. Unification 113
      7.2.7. The most general unifier 114
      7.2.8. An example 115
      7.2.9. Procedural semantics for Horn Clauses 116
   7.3. Main ideas of logic programming 117
      7.3.1. Logic programs as executable specifications 117
      7.3.2. Reversibility of predicates 119
      7.3.3. Logic as a representation paradigm 120

8. PROLOG 123
   8.1. Features of PROLOG 124
   8.2. PROLOG for logic programming 125
   8.3. PROLOG as a deductive database 127
   8.4. Non-logical features of PROLOG 129
      8.4.1. Features for input and output 130
      8.4.2. Features to affect the database 132
      8.4.3. Features to affect control 132
      8.4.4. Implementation of negation as failure 134
      8.4.5. Example of use of non-logical features 135

Contents

	8.4.6. Arithmetic in PROLOG	136
8.5.	PROLOG as an AI programming language	137
8.6.	Summary	138

9. Expert Systems — 141
   9.1. Why expert systems? — 141
   9.2. What is an expert? — 143
   9.3. What is an expert system? — 144
       9.3.1. Expert systems have domains of expertise — 145
       9.3.2. Expert systems are interactive — 145
       9.3.3. Use of heuristics — 145
       9.3.4. Uncertain and incomplete information — 146
       9.3.5. Other possible features — 146
   9.4. Basic expert systems components — 148
       9.4.1. Knowledge base — 149
       9.4.2. Inference engine — 148
       9.4.3. User interface — 149
       9.4.4. Some points about expert systems — 151
   9.5. Early expert systems — 152
       9.5.1. MYCIN — 153
       9.5.2. XCON — 153
       9.5.3. PROSPECTOR — 155
       9.5.4. INTERNIST — 156
       9.5.5. Summary of early expert systems — 157
   9.6. Expert system shells — 157
       9.6.1. Impact of shells — 158
   9.7. Typical facilities of an expert system shell — 160
       9.7.1. User interface — 160
       9.7.2. Knowledge base — 161
       9.7.3. Inference engine — 163
       9.7.4. Escape to underlying system — 163
   9.8. Trends in expert systems — 164
       9.8.1. What kind of tasks can we use expert systems for? — 165
       9.8.2. What are the advantages of expert systems? — 166
   9.9. Examples of current expert systems — 167
       9.9.1. Latent damage advisor — 167
       9.9.2. Pensions advice — 168

10. Some Issues in Knowledge Representation — 171
    10.1. Similarities between the paradigms — 171
    10.2. Expressiveness of Horn Clauses — 172
        10.2.1. Negation in logic programming — 174
        10.2.2. The closed world assumption — 175
        10.2.3. Limitations of negation as failure — 177
        10.2.4. Other treatments of negation — 178
        10.2.5. Negation in other paradigms — 180
    10.3. Non-monotonic reasoning — 182
        10.3.1. Non-monotonicity and frame representations — 184
        10.3.2. Plausible reasoning — 185
        10.3.3. Formal models of default reasoning — 186

		Contents	xi

	10.3.4. Truth maintenance	188
	10.3.5. Non-monotonicity and PROLOG	189
10.4.	Inexact reasoning and rule-based systems	190
	10.4.1. Sources of uncertainty	191
	10.4.2. Heuristics in backgammon	191
	10.4.3. Treatments of uncertainty	193
	10.4.4. Use of probabilities	194
	10.4.5. Certainty factors	196
	10.4.6. Fuzzy logic	198
	10.4.7. Qualitative representation of uncertainty	201
	10.4.8. Summary	202
10.5.	Representation of control knowledge	202
	10.5.1. Representation of control knowledge in the MECHO system	204
10.6.	Time	205
10.7.	Model-based representation	206
	10.7.1. Advantages of model-based representations	208
10.8.	Conclusion	209

Bibliography 211

Index 215

1
Introduction to Artificial Intelligence and Knowledge-Based Systems

1.1. Artificial and intelligent

The term "artificial intelligence"—almost always abbreviated to "AI"—is convenient both because it gives an intuitive characterisation of what this area of computer science is about, and because it conveys an air of excitement which stimulates enthusiasm. But in other ways it is something of a liability, in that it often arouses unjustifiable expectations, and because any precise definition is elusive. In this section I want to examine some of the various things that people have meant by the term. The purpose is not to give a tight definition, but rather to create an awareness of the issues involved. So let us begin by asking what would be expected from an artificially intelligent system. Clearly, if tautologically, we would expect it to be both artificial and intelligent. The "artificial" part gives little problem. A thing is artificial in the sense under discussion simply if it is made by man, although perhaps there are also some connotations of "being made in imitation"—the sense in which "artificial" is opposed to "real", as in "an artificial smile"—so that we could see an AI system as one which attempts to imitate natural intelligence. But in the context of computer science we may well be satisfied with the idea of artificial as simply being a matter of being embodied in a computer system—which is clearly made by man, and which may be intended to imitate human behaviour. Thus an AI system can, for our purposes, be taken to mean an intelligent computer system.

Now some philosophers have argued that the idea of an intelligent computer system is simply absurd, and that it can be shown that no computer system could be intelligent. The champion of this stance in the early years of AI was Hubert Dreyfus [1], while currently it is John Searle [2]. Their arguments turn on their views on philosophy of mind and are proper to philosophy rather than computer science. Searle's main argument derives from a "thought experiment" in which a person is shut in a room and given a set of rules, written in English, to manipulate Chinese characters. He is then presented with sequences of Chinese characters and follows the rules so as to produce further sequences of Chinese characters. To those outside the room it might appear that they were asking questions in Chinese, and the man in the room was replying in Chinese and so they might conclude that he understood Chinese, whereas it is clear that he does not: he is merely following rules in an unthinking fashion. Searle claims that there is no understanding of Chinese here, and no intelligence in the replies, and then contends that the same would be true for any computer system embodying a formal program. What is missing is understanding. Stated in its sharpest form, Searle's claim is that even if the behaviour of a machine acting in accordance with a formal program was indistinguishable from that of a person, we should not ascribe intelligence to it if we knew that it was a machine, although we might mistakenly do so if we were unaware that it was a machine. The variety of responses to Searle [2] indicate, however, that workers in AI would count such a system as intelligent. The disagreement really turns on what is meant by "intelligence": for Searle it is inextricably bound up with the notion of "intentionality", which goes beyond observable behaviour. This is not the place to discuss intentionality, which is a rather unclear notion in philosophy; rather we should note that the possession of intelligence is not a straightforward thing, and that should we wish to characterise our work as the building of intelligent computer systems, we need to be able to say what we mean by this, in the context of computer science. Even if we were to accept the arguments of these philosophers it would show no more than that "artificial intelligence" is something of a misnomer, because there would remain the body of work done under the AI banner, and this work would remain worthy of study. What we need to do, therefore, is find what is meant by "intelligent" in this specialised context.

1.1.1. Popular views of intelligent machines

It is, however, worth noting in passing that the popular imagination has never experienced difficulty in attributing intelligence to machines. In the coffee houses of eighteenth century France a gambler made a fair sum of money by inducing people to play chess against what he claimed was a chess playing automaton, known as the Turk. That it was a fraud consisting of a dwarf in a box was discovered by inspection rather than reflection on the absurdity of an intelligent machine. (In those days it was considered necessary to possess intelligence to play chess.)

Similarly, when Babbage developed his Analytical Engine, in the early nineteenth century, so eager were people to ascribe intelligence to it that disclaimers were felt necessary. For example we find Ada, Lady Lovelace, Babbage's associate, writing

> [It] has no pretensions whatever to originate anything. It can do whatever we know how to order it to perform. It can follow analysis, but it has no power of originating analytical relations or truths. Its province is to assist us in making available what we are already acquainted with. [3]

If we consider the public conception of computers in the twentieth century we find a similar phenomenon. The public press often refers to computers as "electronic brains" and people happily watch films like 2001 in which the computer HAL is accepted as more or less the only intelligent being around. Computer scientists, in contrast, used to be at pains to emphasise that the computers are really pretty dumb, relying on pure speed of computation, and following the orders of their programmers to the letter, no matter how silly the results. The issue of bills for £0.00 is a recurring source of entertainment for the popular media as an example of the alleged stupidity of computers. On this view computers are seen as "fast idiots" rather than "electronic brains".

The growth of AI has confused these traditional terms of the debate because we find some computer scientists wishing to ascribe intelligence to their systems, while, of course, retaining the conventional computer science view for conventional systems. The whole problem really turns on the definition of "intelligence". I would contend that the underlying definitions used by philosophers like Dreyfus and Searle, which derive from particular philosophies of mind, are not really in correspondence with the popular notion. But neither is the extreme opposite used by the AI scientist who said that his thermostat was intelligent because it knew when the room was hot and turned the temperature down. There does seem to be a need to point to some requirement for the use of what we might, at the risk of begging the question, term "higher mental processes" in the task before we wish to ascribe intelligence to its performance. Behaviour of the sort that is no more than a pre-determined response to a stimulus, as is the case with the thermostat, is not enough. What we want to see, and what those striving to build AI systems want to see, is the computer "originating" something. We want to see it telling us something we did not know and would not have thought of, or worked out from the information supplied to the system, before we feel that we should ascribe intelligence to it.

1.1.2. Human versus machine intelligence

In discussing the notion of intelligence as applied to computer systems, however, we should be aware that what counts as an intelligent computer system is not the same as what we would count as intelligent for a human being. In particular, if we see a child capable of accurately performing complex pieces of arithmetic, we should be likely to think him clever, whereas a child who was unable to recognise his family and friends by looking at them would be thought very dull indeed. Quite the reverse is true of computers: the ability to do arithmetic is so taken for granted that this ability counts for nothing in assessing the claims to intelligence of a program, whereas the recognition of faces, and even simple language understanding, is considered to be firmly within the field of AI. This view makes intelligence relative to a norm: every sort of thing, be it a person or a machine, has a set of competences which is expected of it, and only where it goes beyond these do we start to talk of intelligence.

This is the kind of consideration that has led to some popular definitions of AI; such as that offered by Rich in her book *Artificial Intelligence*, that

> AI is the study of how to make computers do things, at which, at the moment, people are better. [4]

On this view, at present people are better at recognising faces, while computers are better at arithmetic; so the former is AI and the latter is not.

1.1.3. Turing test

The definition given by Rich is still pretty vague, however, so if we are to attempt to decide in any objective sense whether or not a system is intelligent, we must propose some precise criterion for the presence or absence of intelligence. The most frequently cited test of this sort is the *Turing test* first proposed by Alan Turing [5]. In this test a person converses with another person and the putatively intelligent system over two teletypes. He asks a series of questions and is supposed to decide, on the basis of the answers, which set of responses come from the person and which from the computer system. If he is unable to decide, then the system passes the Turing test and may be considered intelligent. The test was based on a party game where a person conversed through another person with two people of different sex and was supposed to determine which was the man and which was the woman from their responses.

Despite its appeal, the Turing test runs into two kinds of difficulty. First it sets an impossibly high standard. If the user is permitted to ask any questions whatsoever, and Turing's original statement of the test makes it clear that this is the intention, the test is unlikely to be achieved in the next couple of generations. All successful AI programs, even those whose success is modest, have achieved such success as they have achieved only by focusing on a limited domain. Even to parse an indefinite range of questions is well beyond our current capabilities, let alone to use the sophisticated answering strategies that Turing seems to have in mind. As an example of this game-playing strategy, in his original article an illustrative test asks a complicated multiplication question and the machine not only pauses before answering, but gets it wrong as well.

The second kind of difficulty is that the test is, if applied with sufficient generosity to allow for the possibility of success, too easy. For programs that have been written have managed to pass, admittedly under propitious conditions, the Turing test, even though no one would even begin to claim that the programs in question manifested intelligence in any real sense. ELIZA and PARRY are the classic examples [6]. The problem here is that the user of the system attempts to supply meaning to the output, and can place an intelligent construction on behaviour which is in fact purely mechanical. Thus the Turing test, though an interesting idea and often used as a discussion point, fails to provide us with a practical touchstone.

1.1.4. Typical AI applications

Given the inconclusive nature of the above discussion we might well feel that it would be better to define AI not by the attempt to create intelligence in a computer system, but rather by reference to the kinds of activity that are well accepted as being within the AI domain. This way we do not have to define "intelligence" at all. On this view, we do not take the constituent terms of artificial intelligence seriously at all, but use it as an appealing label for a group of applications. Thus we could point to a long list of applications that would be accepted as within this area of computer science. We could suggest game playing, problem solving, solution of IQ tests, language translation, theorem proving, natural language understanding, planning, vision, speech recognition, robotics and expert systems as a reasonable, although not exhaustive, list. This approach is consistent with the one taken by Rich when she gave the definition cited above. Her definition has the advantage of picking out an identifiable body of work, much as given above, but it suffers from two disadvantages. First, and most obvious, it is a definition which will be constantly shifting. Any successful AI program will have the effect, by virtue of its very success, of ceasing to be an AI program. Perhaps this does not really matter, given the slow rate of progress in AI, but it seems to make AI a fairly unrewarding field in which to work. It is really rather too close to the cynic's definition of AI, that "AI is what we don't know how to do yet", for complete comfort.

A deeper criticism, which applies to any attempt to characterise AI by reference to a list of applications, derives from the suggestion that "it ain't what you do it's the way that you do it". Chess programs are a case in point here. Game playing in general, and chess in particular, have held attractions for workers in AI, perhaps because chess is seen as the intellectual game *par excellence*. Because of limitations in the power and capacity of early

6 Knowledge Representation

computers, a brute force search strategy was impossible in the early days and instead attempts were made to find a way of evaluating positions and moves on a basis other than pure look-ahead. Such programs played chess rather poorly. Current chess computers play chess rather well, but this has not been because of the increased sophistication of the evaluation function, which would represent the incorporation of chess-playing knowledge into the program, but because hardware developments have increased the look-ahead capability to an extent which renders the use of judgement of largely decorative importance. Success in chess has been attained, not by creating chess-playing intelligence, but by having computers capable, by virtue of their speed, of playing competent chess without the need for chess-playing intelligence, by simply examining a large number of positions and applying a fairly crude evaluation function to them. So chess, it can be argued, has become successful by ceasing to be AI; but this is because of a methodological change, so we are not saying that chess has ceased to be AI simply because it is successful. Also, chess programs would remain AI programs on Rich's definition, irrespective of the methods used, since people remain, at the highest level, better at chess that computers.

The flaw in the approach of definition of the area by respect to applications is essentially that it overlooks the necessity of not just doing a task which requires intelligence, but of doing it in an intelligent way. It is this requirement, rather than doubts about its performance, which underlies our refusal to credit intelligence to a computer system solely on the basis of its brute force application of a searching algorithm.

1.1.5. AI and understanding the human mind

Following this last line of thought, one can define AI from the perspective of cognitive psychology. From this point of view, method is all important, and the nature of the application and achievement become entirely secondary. The primary interest of cognitive psychologists is in the human mind, and the construction of AI programs is pursued for the insight it affords into the working of the human mind. The idea is that the computer program can be seen as a model of actual human reasoning processes, and the success of the program as a vindication of the use of the model. In practice the areas explored are much the same, but the criteria for evaluating a program are very different; there is no point in enhancing the performance of the program at the expense of fidelity to the psychological model being explored.

This is the standpoint of Charniak and McDermott [7] when they offer as a definition of AI

> AI is the study of mental faculties through the use of computational models.

This definition, embodying the cognitive science standpoint, seems to make computer science auxiliary to AI, rather than seeing AI as part of computer science, and as a consequence it fails to cover a number of developments motivated from a computer science standpoint that we might want to include as AI, such as automated theorem proving, where automated methods are quite unlike human methods. It remains, however, a major motivating force for a good deal of AI research.

1.1.6. Techniques used in AI

The final way of defining the area of AI that I shall consider, and the one that I shall follow most closely in this book, is one which seeks to define the area by reference to the techniques employed. Although there are a variety of techniques which can properly be seen as AI techniques, such as neural networks and work based on distributed parallel processing, two techniques are of particular concern in this book: the first is the technique of search, especially search techniques involving heuristics, and the second is the idea of separating a knowledge base from a program which operates on that knowledge base. Search is an important AI topic, and we shall be looking at it in Chapter 4, but the second technique, which views an AI program as a knowledge base together with a set of methods for manipulating it, is what we are going to be concerned with throughout the book. If we assign this key role to the knowledge base, the form of representation assumes a crucial importance in the construction of our programs.

From this perspective an AI program is simply one which works by having a declarative representation of some body of knowledge relevant to a problem, and by manipulating this knowledge. This is intended to contrast with conventional programs which work by expressing algorithms in some imperative language.

1.1.7. Different motivations for AI

The multiplicity of definitions of AI is a reflection of the different motivations of those who work within the area. At one extreme is the motivation of Charniak and McDermott, who say "the ultimate goal of AI research is to build a person, or, more humbly, an animal" [8]. At the other is the Rich approach, which justifies the study of AI by saying that its techniques can help us to build programs to do things we could not build programs to do before. Additionally there is the view which sees AI, and its applications, primarily as a stimulus to the development of interesting computer techniques. All of these motivations are perfectly legitimate; but the kind of systems which are produced, and the claims made for them, and the assessment of them, will depend crucially on the standpoint adopted.

1.1.8. Knowledge-based systems

Another term, which is especially associated with the techniques-orientated view which gives a key role to the knowledge base, and which is often used today, is *knowledge-based system*. This term is largely used by those who wish to avoid philosophical debate about what they are doing arising from the connotations implied by the use of the term "intelligence". The use of this term also ought to imply a commitment to a certain subset of AI techniques and tends too to be associated with exploitation of these techniques in commercially viable applications.

Finally I should mention the term *expert system*. Such systems are a subset of knowledge-based systems, which exhibit a number of characteristics, loosely derived from the behaviour attributed to people who are experts in a specialist domain. The characteristics of expert systems will be discussed in full in Chapter 9.

To conclude this chapter I should like to give a quotation from the invitation to a ten-man two-month workshop that was held in Dartmouth, USA, in 1956. This conference really marks the beginnings of AI as a study, since it represents the first significant time the term "artificial intelligence" was seen in print:

> The study is to proceed on the basis of the conjecture that every aspect of learning or any other feature of intelligence can in principle be so precisely described that a machine can be made to stimulate it.

This conjecture is at the bottom of nearly all work in AI, and in particular the study of knowledge representation is an attempt to show it to be true; it is a search for the ways in which this precise description of features of intelligence can be given. At the time of the conference, hopes were high, but little, apart from the founding of AI, was achieved there and the truth of the conjecture remains a matter of debate. Perhaps it should be seen as an aspiration rather than a hypothesis, and it is instructive to see the subsequent development of AI as various strainings towards that ultimate (perhaps unachievable) goal.

References

1. Dreyfus, H., *What Computers Can't Do—A Critique of Artificial Intelligence*, Harper and Row, New York, 1972.
2. Searle's argument turns on his example of the Chinese Room, which has attained a certain degree of notoriety. The original paper is in "Minds brains and

programs", *Behavioural and Brain Sciences 3*, pp. 417–424, 1980. Some of the hostility it aroused among AI people can be seen from the peer group commentary published in the same volume.
3. Lovelace, A., "Notes upon L.F. Menabrea's sketch of the analytical engine of Charles Babbage", in *Charles Babbage and His Calculating Engines* (ed. P. Morrison and E. Morrison), Dover, New York, 1961, p. 284.
4. Rich, E., *Artificial Intelligence*, McGraw-Hill, Singapore, 1983, p. 1.
5. Turing, A., "Computing machinery and intelligence", in *Computers and Thought* (ed. E.A. Feigenbaum and J. Feldman), McGraw-Hill, New York, 1963.
6. ELIZA is a program designed to simulate an analytic psychoanalyst developed by Joseph Weizenbaum, and PARRY a program to simulate a paranoid, developed by K.M. Colby. Both rely on a simplistic matching of an input to produce a response. For a discussion of them in relation to the Turing test, see Boden, M., *Artificial Intelligence and Natural Man*, Harvester Press, 1977, chapter 5.
7. Charniak, E. and McDermott, D., *Introduction to Artificial Intelligence*, Addison-Wesley, Reading, Mass., 1985, p. 6.
8. Charniak and McDermott, *op. cit.*, p. 7.

2
Introduction to Knowledge Representation

2.1. What is knowledge representation?

When we come to write AI programs we find that there are a number of different kinds of things that we want our program to "know". It will need to know a large number of facts, and to know about the processes which cause those facts to change. It will need to know about the objects which those facts concern, and the relationships in which those objects stand to one another. If it is intended to solve problems of some sort, it will need to know about the problem and what will count as a solution to the problem, together with some strategies for solving such problems. This list, while not exhaustive, represents quite a diverse collection. The study of knowledge representation concerns the ways in which we might go about encoding them in a computer program.

If we take the above list as exemplifying what we mean by "knowledge", we may briefly consider what we mean by "representation". We may offer the following definition:

> a set of syntactic and semantic conventions that makes it possible to describe things.

By *syntax* we mean that which specifies a set of rules for combining symbols so as to form valid expressions. By *semantics* we mean the specification of how such expressions are to be interpreted.

It may be helpful to point to some examples of knowledge representations with which we are all familiar. Maps are a good example. There we have a relatively widely understood set of conventions for what symbols can be used and how they are to be interpreted when attempting to translate from the map to the geographical situation it describes. Another example is provided by the various notations used to describe the positions in a game of chess. Not only are the diagrams we see in newspaper chess columns representations of the state of a game, but so too are the board and pieces used by people when they actually play. Finally we can see natural languages such as

2.1.1. Purpose of knowledge representation

English as the most expressive and powerful knowledge representations available to us.

At this point it is worth considering an example to illustrate what we can hope to get out of a successful knowledge representation. Suppose we play the following game: there are nine counters, each with a number from 1 to 9. The players take it in turn to pick a counter, and the first player to hold three counters which have numbers adding up to 15 is the winner. Let us consider for a moment how we would represent the state of the game.

One way would be simply to keep three lists, one for each of the collections of counters held by each of the two players, and one for the counters yet to be chosen. But seeing whether we have a winning move involves taking each combination of two counters from our own list and seeing whether this has a sum which is 15 minus the value of one of the available counters in the unchosen list, and the representation is even harder to exploit when trying to see which counter we should choose in a situation where no immediate win is possible. It is worth trying to play the game a couple of times to see just how difficult it is to keep a track of the state of play and the opportunities available.

But now suppose instead we form a magic square (as in Fig. 2.1) with each line summing to 15, and record the first player's moves by putting an X through the appropriate number, and the second player's moves by drawing a circle around the appropriate number, unchosen numbers being those with neither indication. Now we have transformed the game into noughts and crosses. This makes it far easier for us to play, since we can now see winning combinations at a glance, and apply the standard noughts and crosses strategy with which we are familiar. Now what was quite a difficult game is trivially easy.

From this example we can see two of the things we are trying to do by our choice of knowledge representation. We want to make the processing we have to do as easy as possible, and we want to use the representation to help us to map novel problems on to problems which are familiar and well understood.

One final thing worth noting about the example: the magic-square representation made it easier for us, but if we were trying to write a computer program to play the game, this representation might not be so good, because "seeing winning lines at a glance" is a human rather than a machine-orientated skill. In fact we might prefer to use the original form of representation as a machine-orientated representation for noughts and crosses.

To summarise, we want a knowledge representation which is right for the

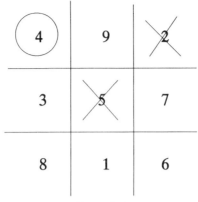

Figure 2.1

domain (which is the knowledge to be represented) and right for the task (which is what needs to be done with the knowledge) and right for the user of the knowledge, whether human or machine, which must actually perform the manipulations.

2.2. Criteria of adequacy

The above example gives us some notion of what we are trying to achieve with a knowledge representation. Often, however, there will be a number of competing candidate representations, and we will need to select one of them as the most appropriate. For this reason, we need to consider the criteria which we can use to assess the various possibilities. Some of these criteria will relate to things which the representation must have if it is to do what is required of it, and others will relate to desirable aspects of the representation. The former group of criteria may be termed *criteria of adequacy* and if the representation fails on one of these respects, whatever virtues it may possess when considered on other grounds, will be of no avail, since we will not be able to represent the knowledge that we need. The first three criteria of adequacy I shall discuss derive from Hayes and McCarthy's paper "Some philosophical problems from the standpoint of artificial intelligence" [1].

2.2.1. Metaphysical adequacy

Firstly our representation will need to be *metaphysically adequate*. By this we mean that it must be the case that there is no contradiction between the facts that we wish to represent and our representation of them. To use Hayes' and McCarthy's example, we could not produce an adequate representation of physical objects in the world if we tried to represent them as a collection of non-interacting particles. Since objects do in fact interact with one another there is a fundamental contradiction between represented and representation which renders the representation metaphysically inadequate. Note that this criterion does not determine the representation we must use: representation of the world either as a collection of particles which interact through forces between each pair of particles, or as a giant quantum mechanical wave function, would be metaphysically adequate, in that there is no such contradiction. At this level the representations are mainly useful for constructing general theories.

For another example consider the game above. A representation whose syntax allowed a number to be both crossed through and circled would be metaphysically inadequate, since it would conflict with the fact that once a counter has been chosen by one player it is not available to the other player.

2.2.2. Epistemic adequacy

Of course, metaphysical adequacy is not enough, since we need to make the step of deriving observable consequences from our general theories. This introduces the need for *epistemic adequacy*, which is to say that the representation must provide us with the ability to express the facts that we wish to express. Neither of the metaphysically adequate representations mentioned above is able to express such commonplace facts as that the President of the United States of America lives in the White House, or that Margaret Thatcher is currently Prime Minister of the United Kingdom, and we are very likely to want to include these kinds of facts in our knowledge base. Thus to be epistemically adequate for this purpose a representation must provide ways of referring to particular individuals and their relations and attributes.

In the above game, epistemic adequacy explains why we need two different symbols to represent the counters chosen by the different players: otherwise we could only represent that a counter had been chosen, not which player had chosen it.

2.2.3. Heuristic adequacy

This third criterion was somewhat tentatively proposed by Hayes and McCarthy, and was meant to suggest the need for the representation to be itself capable of expressing the reasoning that is gone through in solving a problem. This is perhaps the most difficult to fulfil, and it is not clear that any of the paradigms I shall consider later satisfy it completely. It would be a definite requirement if we wished to build systems with a certain degree of "self-consciousness", capable of reflecting on their own reasoning, but it is not clear that it is required to do any AI work at all, since this sort of knowledge may be represented differently, or may even be entirely implicit in the program code. So representations which are perfectly acceptable in other respects may not be heuristically adequate; this will, of course, limit what can be done with them, but does not make them entirely unusable.

In the example the representation is not heuristically adequate: it only represents the state of the game, and a quite different representation would be required to represent knowledge about how to play the game.

2.2.4. Computational tractability

The criteria discussed above would hold good for the representation of knowledge for any purpose whatsoever. We are, however, interested in representing knowledge for a particular purpose, namely so as to be able to incorporate it within a computer system. Therefore we need to take account of this, and impose another criterion of adequacy, that it should be *computationally tractable*. This requires that we be able to manipulate the representation efficiently within a computer system. So, whilst, to take an obvious case, the English language as used by native speakers scores well on the above criteria, in would be useless as a computer-orientated knowledge representation, since it is not (at least currently) possible to use it in computer systems.

What is computationally acceptable may change as advances are made in computer hardware, and in methods for manipulating representations. It is always true, however, that when we come to build a system we are faced with a given set of hardware and efficiency constraints, so that in a practical situation we may need to compromise some of the theoretical elegance of our representation in order to meet this criterion. After all, if we cannot compute with the representation, we will simply have no program.

2.3. Expressiveness

The above criteria all represent conditions that a representation must fulfil if it is to be used at all. There are in addition a number of other considerations that can enable us to choose amongst different representations. These considerations represent desirable features, but are far less clear cut than the other criteria, as different representations will score differently when considered with respect to different applications, and to different people. These considerations may be summarised as *expressiveness*, by which I mean that not only can one say what one means, but also that one can say it clearly and without ambiguity.

2.3.1. Lack of ambiguity

The requirement to be unambiguous means that every valid expression in the representation should have one and only one interpretation. This is essentially the requirement that the semantics of the representation be well defined. Natural languages such as English, while scoring well on other non-computational criteria, tend to fall down when it comes to being unambiguous. Thus the sentence "every student uses a computer" could mean either that there is some single computer used by every student (as when they log on to a central mainframe), or that every student uses a computer, but not necessarily the same one (as when they each have their own personal computer). When we speak English we rely on pragmatic features such as context and background information to resolve such semantic ambiguities, but this context cannot be assumed to exist for a knowledge representation. The development of logic was to a large extent motivated by the need to disambiguate sentences of natural English.

2.3.2. Clarity

The need for clarity arises because although we intend the representation to be executed within a computer system, we none the less need to construct the knowledge base, which will involve us in writing the knowledge we wish the system to manipulate in the chosen representation, debugging the knowledge base when the program performs in an unexpected manner, and probably discussing the represented knowledge with someone who has more knowledge of the field than we do. All of this means that the representation must be amenable to understanding by people, even those who may not be entirely immersed in the particular representation formalism. It is an additional benefit if, when the knowledge is represented, our understanding is clarified and increased, rather than being clouded. In short, what we are seeking here is a representation which is such that if we present an expert in the field with the knowledge base, he should be able to make constructive comments, and correct errors, and perhaps even become aware of knowledge that he did not know that he possessed.

2.3.3. Uniformity

At the start of this chapter I listed a number of different types of knowledge that we might want to represent in an AI system. It is obviously an advantage if our chosen form of knowledge representation is able to handle all the different types of knowledge that we wish to deal with. Thus the more kinds of knowledge that a representation can cope with, the better we shall regard that representation. Often, however, it will be the case that we cannot fix on a single all-embracing representation. In such cases the important thing is to be consistent in our choices, so that all knowledge of a given type is represented in the same way, and that we have principled reasons for choosing how to represent a particular item of knowledge. The criterion of uniformity is thus suggesting that in any given system, the manner of representing a given item of knowledge should not be an arbitrary choice.

2.3.4. Notational convenience

A more subjective, but by no means unimportant, consideration is that of the convenience of the notation for representation. Convenience may relate to the knowledge we are trying to represent, in that some kinds of knowledge may fit more naturally some representations rather than others. Also it may be a matter of personal preference on the part of the person building the system or supplying the knowledge. Some people feel more comfortable with certain formalisms than others. It is important to choose one which all concerned like and like using.

2.3.5. Relevance

When assessing knowledge representations it is important to remember that we are typically representing knowledge for a particular purpose. This means that our assessment of a representation cannot be absolute, but will be relative to the task we are trying to perform. Given also that a gain in the expressive power of a representation is usually bought at the expense of making the representation more complex and difficult to manipulate, we need to be sure that the extra expressiveness is required by the task. In Chapter 3 we will be looking at logic and we will meet both the propositional and the predicate calculus. The latter is more expressive than the former, but the former is sufficient for many purposes. Thus when applying the criteria outlined above, the reason for representing the knowledge needs to be kept in mind, so that any failures of expressiveness which dictate a complication of the representation can be seen to be really needed.

2.3.6. Declarativeness

Before leaving the topic of desirable features of representations, we should consider the important notion of *declarativeness*. A declarative representation is one in which the meanings of the statements are independent of the use made of them, and which has the property of *referential transparency*. A representation is referentially transparent if equivalent expressions can always be substituted for one another whilst preserving the truth value of the statements in which they occur. Mathematics and formal logic are referentially transparent, whereas conventional programming languages are not. That this property does not hold for conventional imperative programming languages can be seen from considering an example such as the following BASIC program for calculating the factorial of a number.

```
30  LET T = 1
40  FOR K FROM 1 TO N
50  LET T = T*K
60  NEXT K
70  PRINT "FACTORIAL";N;"IS";T
80  STOP
```

Provided N is set to the required value before the loop is entered, this will calculate the value of factorial N. But the statements which make up the program do not have a meaning independent of the computation of the program: their effect depends on the computation that has occurred before they are executed, and hence on the order in which they occur in the program. Each passage through the loop will increase K by 1 and alter the value of T, and so the effect of line 50 will depend on how far the program has got. Line 30, read declaratively, would suggest that we could substitute 1 for T: but to do so in line 70 would make a nonsense of the program, since the whole point of the program is to change the value of T to the value of factorial N. The process of changing the values of the variables in this manner in the course of executing the program is known as *destructive assignment* and is fundamental to conventional programming languages.

But if our aim is to separate the knowledge represented in a program from the techniques used to manipulate this knowledge, we shall want the knowledge to be represented in a declarative fashion, for we can make no assumptions as to what computation will have taken place before a particular statement in the knowledge base is used. Ideally then, we would want our representation to be referentially transparent, and for the meanings of statements in the representation to be independent of other statements in the knowledge base and of the means of manipulating the knowledge base. In practice, as we shall see, this ideal of declarativeness must often be compromised, but declarativeness is a property to which the representation should aspire.

2.4. Example of the use of these criteria

As an example of how we can use the criteria developed above to assess a knowledge representation, consider the use of Roman numerals to represent numbers for doing arithmetic. In the Roman system of numerals, certain letters represent different quantities: M represents 1000, D represents 500, C represents 100, L represents 50, X represents 10, V represents 5 and I represents 1. Placing a lower symbol after a greater symbol indicates that they are to be summed: thus CL represents 150. Placing a lower symbol before a greater indicates that it is to be subtracted: thus XC represents 90. Now is this a metaphysically adequate representation? In a sense it is, for positive integers, since it is possible to represent any positive integer using the notation. However, the notation strongly suggests that the number sequence is finite, as otherwise we should expect distinct symbols for 5000, 10 000, and so on. Representing a million by a string of 1000 Ms is hardly a practical proposition. In the Middle Ages extensions were used to provide ways of making a number 10 or 1000 times its normal size: these methods merely postponed the size at which the notation becomes impractical, and so do not suggest the possibility of indefinitely large numbers.

If we turn to epistemic adequacy we see three deficiencies. First there is no way of representing zero, and this is important if we are to do arithmetic effectively. Second there is no way of representing non-integer numbers, although we might propose some extensions which could do this, at least for some classes of non-integer numbers. Third there is no concept of negative numbers. Thus the notation as it stands is epistemically inadequate for other than positive integers.

Heuristic adequacy does not really apply here, but if we turn to computational tractability (for human beings), we find that the notation is hard to manipulate. Unlike normal Arabic notation, there is no relation between the length of a string of Roman notation and the size of the number, the most significant digit does not always occur first, and the numbers cannot be neatly aligned in columns for addition. All of these factors make it extremely difficult to calculate in the Roman notation.

To turn to the criteria relating to desirability, we can say that the notation is unambiguous, but it cannot be said to be clear, largely for the same reasons as those given for explaining its unsuitability for computation. Given the defects outlined above it scores badly as to notational convenience. The only place where it is still widely used is in clock making, where many faces are still designed using Roman numerals. In that specialised use, the greater decorative properties of the notation can outweigh the problems, particularly in view of the fact that only the integers from 1 to 12 are required.

2.5. Major paradigms

In the early days of AI, the subject was primarily concerned with search. This remains an important topic, and Chapter 4 of this book will give an overview of the more important ideas relating to it. At that time the importance of knowledge representation as a separate topic was not recognised, and knowledge representation was largely an *ad hoc* affair, with new representations being developed to meet the needs of new projects, and with little attention being paid to formality and the ability to generalise across projects. As time went on, however, a number of leading knowledge representation paradigms started to emerge, and it is with these paradigms that the bulk of the remainder of this book will be concerned. All of the paradigms satisfy the criteria of adequacy outlined above, at least with regard to metaphysical, epistemic and computational adequacy, but there are differences between them which mean that they rate rather differently on the more subjective, or application-orientated criteria. Here I shall simply list them.

Production rules are the representation of knowledge as a set of condition action pairs. The paradigm owes its popularity largely to work on expert systems such as MYCIN and R1.

Semantic networks and *frames* are two of the types of structured object representation that were first developed to meet the demands of story understanding systems.

The logical representation of *first-order predicate calculus* was an obvious candidate for knowledge representation, since it was developed to provide a clear and unambiguous way of representing natural-language statements. Originally it seemed to fall foul of the criterion of computational adequacy, however. This was mitigated by advances in techniques of automated theorem proving, and now it can be seen as computationally tractable, especially if the representation is confined to a subset of predicate calculus, known as Horn Clauses.

Each of these paradigms will be considered in detail in subsequent chapters.

2.6. Manipulation of representations

We can make a broad distinction amongst the things we know between those that we can recall immediately, and those that we know how to work out. Thus I know, as a brute fact, that 6 times 7 is 42, and I can work out the answer to 123456 times 87647. A similar sort of distinction can be made for AI systems as well. Suppose we were interested in family relationships. We could store a table such as

person	father	grandfather
andy	tom	dick
tom	dick	harry
harry	arthur	bill
john	andy	tom
arthur	bill	ethelred

Now we can answer questions about who is a given person's grandfather by looking up the answer in the table. But the third column of the table is redundant, since we need have only two columns although we would then have needed to add another two lines to record the fathers of bill and dick. We could then work out the answer to questions about people's grandfathers on the basis of the information in the first two columns and the general principle that a person's grandfather is his father's father. Similarly if we wanted to answer questions about who is whose child, we could store the information explicitly in another table, or we could augment the table by the principle that a person is the child of some other person if that other person is the person's father.

When constructing the knowledge base for an AI program we will have to make choices as to what we are to store explicitly and what we are to work out when the program is executed, which is to say, to store implicitly. Such decisions are likely to influenced by such considerations as the frequency with which certain information is needed, and the time that it will take to compute the answers where the knowledge is implicit, and the space we have available for the knowledge base. Other things being equal, implicit storage is to be preferred because it involves less work in maintaining the knowledge base and has less danger of becoming inconsistent if the knowledge base is updated incorrectly.

It is therefore a requirement on our knowledge representation that it be possible to express general principles, like "a person's parent's parent is his grandparent", so that we can derive new facts implicit in our explicit facts, and that the representation of the facts also allows us to make these manipulations. I shall conclude this chapter by looking at some of the different ways in which we can derive new information.

2.6.1. Deduction

Deduction is a form of reasoning which uses the principles of logic. The following are examples of deductive arguments

> All men are mortal
> Socrates is a man
> Therefore, Socrates is mortal

> If it is raining then the streets are wet
> It is raining
> Therefore, The streets are wet

In the above the first two sentences are *premises* and the third is a *deductive conclusion* from these premises. In general, given a set of sentences as premises, we may be able to say that some other set of sentences is logically entailed by the first set. These are the deductive consequences of the original set of premises. Deductive consequences have the nice property that if the premises are true, then the deductive consequences must also be true. That is to say deductive inference is *truth preserving*. This property, of course, makes deductive reasoning especially attractive, since it means that we can be just as confident in our conclusions as in our grounds for making them. That is, we can draw all and any inferences licensed by our premises, and be sure that we shall never need to retract them.

2.6.2. Induction

Induction in contrast is a method of reasoning that is not necessarily truth preserving. As examples of inductive reasoning we may say

> This swan is white
> That swan is white
> Every swan I've ever seen is white
> Therefore, All swans are white

Or again,

> The sun has risen every day so far,
> Therefore, The sun will rise tomorrow.

Inductive reasoning is essentially finding some generalisation that describes a current set of observations. It is a major paradigm of reasoning used in science. But inductive conclusions cannot be accepted with certainty, simply on the basis of acceptance of the premises, since some future observations may conflict with the generalisation, which will then need to be refined or abandoned in consequence. Thus in the case of the first example, the discovery of Australia led to black swans being observed, and the necessity to refine the conclusion to something like "All swans native to the Northern hemisphere are white".

Inductive reasoning, although it can, by its nature, only give rise to provisional conclusions, is an important method of reasoning, without which science would be impossible. It is essential for the generation of working hypotheses, but that the status of inductive conclusions is hypothetical cannot be forgotten. There will always, no matter how impressive the volume of evidence, be the possibility of a counter-example being found.

The above discussion applies to induction on observations, and should not be confused with mathematical induction, which is, of course, a sound principle of inference. The difference is that the basis of induction in mathematics is an ordered set, such as the integers, which can be related to one another in a way that empirical observations never can be. Thus in mathematics we can prove something (deductively) for an arbitrary number and its successor, and for a given number, and so conclude that the thing proved holds for all successors of that number. But no such relationship holds for observations; it is neither possible to prove properties of arbitrary observations, nor to reason from one observation to its successor. Thus mathematics provides a very special basis for induction, which makes it sound despite the cogent objections to the general case. The major place of induction in AI is in the field of machine learning.

2.6.3. Abduction

A third style of reasoning is called *abduction*. An abductive argument is something of the form

> If it is raining, then the streets are wet
> The streets are wet
> Therefore, It is raining.

Now this seems a fairly strange way to reason, particularly for any one with a logical training, since it is an example of a well known fallacy, the fallacy of affirming the antecedent. The reason it is fallacious is that there may be many other causes for the wetness of the streets, and we are entitled to draw the conclusion that it is raining only after we have eliminated all the other possibilities. None the less abduction is used in many situations. Consider:

> If a person has a cold, then he has a runny nose
> Jack has a runny nose
> Therefore, Jack has a cold

This is the kind of inference that we make all the time, and it is so useful as to be indispensable. None the less we need to recognise that abductive conclusions are not certain, and we may have "leapt to the wrong conclusion". Jack may turn out to have hay fever. What we can do with abduction is to generate hypotheses and explanations which we must then test by further inquiry or observations which may confirm or overturn them. Abductive reasoning is sometimes called plausible reasoning, since its conclusions are plausible, but no more than that.

The other sphere where abduction is required is in the field of what might be termed practical reasoning. Consider:

> I must be in London by 11.00
> The earliest train is the 8.15
> If I catch the 8.15 I will arrive at 10.25
> If I catch the 9.05 I will arrive at 11.10
> Therefore, I must catch the 8.15

24 Knowledge Representation

Here we have it in our power to realise the truth of the antecedents in the premises, and so we can use abduction to find which one to realise, and so form a plan to achieve our desired ends.

2.6.4. Other methods of reasoning

In addition there are, of course, a number of other ways of reasoning, ranging from such well understood methods as the reasoning used to solve arithmetical problems, to unclear matters such as intuitive reasoning and reasoning by analogy. Some efforts have been made to render such reasoning forms in terms of other reasoning forms, especially deductive reasoning, where the trick is to identify one or more suppressed premises in the argument. Similarly the sort of reasoning in AI systems, such as semantic nets, employs other modes of reasoning which may or may not be capable of being reconstructed as deductive reasoning. This kind of issue will be discussed in later chapters.

Exercises

2.1. (i) Discuss the criteria that can be used to assess methods of knowledge representation.
     (ii) Shepherds used to record the number of sheep in their fold by cutting a notch in a tally stick as each sheep entered. Use the criteria you gave in (i) to assess this as a knowledge representation formalism. Suggest extensions which might help with any difficulties. Consider both the case where they only want to know how many sheep, and in the case where they may also wish to know how many black sheep, how many brindled lambs and other similar information.

2.2. (i) Chess positions are often represented by diagrams such as the one below. There is an alternative notation, known as the Forsythe notation, in which each rank is described in the following way. A number represents that number of consecutive unoccupied squares, a lower case letter a black piece (k = king, q = queen, n = knight etc.) and an upper case letter represents a white piece. Thus the position below would be represented as

   R3KB1R-1PQN1PPP-4PN2-2pP4-Bp6-2qlp3-pblnbppp-r3k1nr

   Each representation needs to be augmented by the additional information as to which side is to move (white to move in the above case). Also it is necessary to be aware that white moves up the board in the diagram, and that, for Forsythe, the first rank described is white's back rank. Why are these extra pieces of information necessary?

   In what way would the Forsythe notation be inadequate were it not case-significant?

   Compare and evaluate the two representations as representations.

   The diagram does, and Forsythe does not, distinguish between white and black squares. Does this matter? If so, why, and if not, why not?

Introduction to Knowledge Representation   25

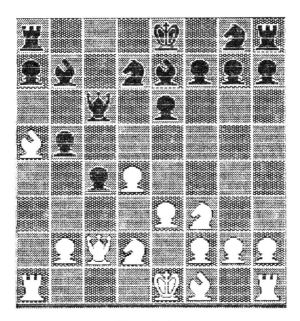

Example 2.2

These notations represent a position. In a game of chess, however, moves are made. How could we augment these representations so as to record a game? The standard method of representing a game records moves, not positions. Why?

(ii) Devise a representation for recording the position of a game of draughts. Does it need to take all 64 squares into account? Would it be suitable for representing the position for computation?

Reference

1. McCarthy, J. and Hayes, P.J., "Some philosophical problems from the standpoint of artificial intelligence", in *Machine Intelligence 4* (ed. B. Meltzer and D. Mitchie), Edinburgh University Press, 1969, pp. 463-502.

3
Logic

This book cannot hope to give a full introduction to logic. Logic is, however, so central to the concerns of knowledge representation, and so useful a tool for describing and understanding the various paradigms, that I have included this chapter which gives a brief outline of some of the basic concepts and terminology which will be used throughout the rest of the book. Those who want to go into greater detail about the subject should work through one of the many excellent introductory books on the subject such as those of Refs 1 or 2.

3.1. Basics

The main concern of logic is with the soundness and unsoundness of arguments, and its goal is to represent an argument in such a way that it will be uncontroversial as to whether that argument is acceptable or not. An argument comprises a set of premises which are known to be true, or accepted as true for the purposes of the argument, and a conclusion which is said to be true on the strength of those premises. The statements used in an argument will have content, but they will also have form. Thus

> A1 Omnium is a Duke; no Dukes are Members of the House of Commons; therefore Omnium is not a Member of the House of Commons

and

> A2 Fido is a dog; no dog can retract its claws; therefore Fido cannot retract its claws

Both are, of course, sound, and though they differ greatly in their content, share a common form. The soundness of these arguments, moreover, follows from their form, and is entirely independent of their content: any argument of this form, no matter what it is about, will also be sound. The idea underlying logic is to represent arguments by abstracting away from their content to their form, and to present a means of discovering whether a given argument is sound in virtue *of its form alone*. Logic is thus primarily concerned with the form of arguments, rather than with particular arguments.

Logic has a lengthy history: usually taken to begin with Aristotle's work with syllogisms. A syllogism is a particular form of argument consisting of two premises and a conclusion. A1 and A2 represent syllogisms as does

>A3 All elephants have four legs; Fido has four legs; therefore, Fido is an elephant.

A vital difference is, of course, that A1 and A2 are sound and A3 is not. Aristotle's work consisted of a codification of the various logical forms that syllogisms could exhibit, in terms of the universal terms and individual terms within them, and saying which forms were sound and which were not.

The syllogism was useful, but extremely limited, in that many arguments cannot be cast into this form. Modern logic really owes its development to Frege, who produced the predicate calculus, and Frege's work forms the basis for contemporary logic. Before proceeding to the predicate calculus, however, it is essential fully to understand a rather simpler logic, the *propositional calculus*.

3.2. Basics of propositional calculus

The propositional calculus is concerned with propositions. A proposition is expressed by a sentence which says something that is either true or false. Many of the sentences we use do this; such as that John is fat, or that the White House is in Washington. Many other sentences, however, do not: questions and exclamations, for example. In some cases whether or not a sentence expresses a proposition or not is a matter of controversy; some hold that value judgements, such as "killing people is wrong" or "Picasso is a great artist", cannot be said to be true or false, for example. This is not the concern of the propositional calculus, however; it deals only with sentences that do express propositions, and if value judgements do not, then it does not deal with them. In what is termed classical logic, moreover, propositions have one, and only one, truth value; they are either true or they are false. Non-classical logics tend to alter this fundamental assumption, either by allowing more truth values, or by allowing some propositions to lack truth values. We will restrict ourselves to classical logic in what follows. A final point about propositions: they possess their truth values timelessly. Thus if a proposition is true, it always has been and always will be true. This can give rise to difficulties: "I am fat", expresses a proposition, but whether it makes a true statement or not depends on who utters it, and on the time at which they do so. This does not mean that the truth value of the proposition expressed changes; the proposition expressed is found by dereferencing the utterance, so as to make explicit the referent of the utterance, and the time of utterance. Thus the truth value of "I am fat" does not change, but the proposition expressed by the sentence does, according to the speaker and the time of utterance. In the example above, therefore "I am fat" needs to be understood as "Trevor Bench-Capon is fat as of 7 September 1989".

3.2.1. Notation of propositional calculus

The propositional calculus uses two types of symbol in its notation. First it requires symbols to stand for propositions. These are called propositional variables, and are upper-case letters, usually starting with P. These propositional variables stand for anything whatsoever that can be ascribed a truth value, irrespective of content or internal form. In order to represent complex arguments, however, we need also symbols to represent the ways in which propositions are connected within arguments. Propositional calculus uses the following operators:

- $-$, which represents *negation*: thus $-P$ may be read as "not P"

- $\rightarrow$, which represents *material implication*: this is often read as "if . . . then", so that $P \rightarrow Q$ is read as "if P then Q". This can, however, be misleading, as we shall see

- &, which represents *conjunction*: thus P & Q may be read as "P and Q"

- $\vee$, which represents *disjunction*: thus $P \vee Q$ may be read as "P or Q". Note that the disjunction is *inclusive*, so that P and Q may both be true

- $\leftrightarrow$, which represents *equivalence*: for convenience $P \leftrightarrow Q$ may be read as "P if, and only if, Q", although this may again be misleading on occasion

Using these operators, we may produce statements of arbitrary complexity. Thus

$$P \leftrightarrow Q \vee R \, \& -S \rightarrow P$$

is a valid expression of propositional calculus. The operators have precedence rules which tell how they are to be read, but for convenience brackets are often used to make this clear. The above expression could thus be more clearly written as

$$(((P \leftrightarrow Q) \vee (R \,\&\, -S)) \rightarrow P)$$

One final symbol should be introduced here. In representing arguments we may wish to distinguish between the premises and the conclusion. For this purpose the symbol "$\vdash$" is used. Thus we can read $P \rightarrow Q$, $P \vdash Q$ as "Given if P then Q and P, we may conclude that Q".

3.2.2. Truth tables

Although readings of the various logical operators were given above, these were for convenience only. The English language is inherently ambiguous, and several of the readings contain connotations and typical implications which may or may not hold good for a particular use of the expression. This will not do for logic, and thus logic defines the operators in a way which is totally unambiguous. It does this by making the truth value of a complex expression formed from propositional variables and operators a function of the truth values of its constituent propositions. These functions are conveniently expressed by the means of truth tables, which list the possible truth values of the constituents and the value of the complex proposition. The truth tables for the connectives are:

P	Q	$-P$	$P \,\&\, Q$	$P \vee Q$	$P \rightarrow Q$	$P \leftrightarrow Q$
T	T	F	T	T	T	T
T	F	F	F	T	F	F
F	T	T	F	T	T	F
F	F	T	F	F	T	T

These truth tables express the real meanings of the operators, and where there is a conflict with the English reading, the latter should be ignored. Negation is relatively uncontroversial, but with conjunction there is often some implication of a temporal relation between the two propositions. Thus "I gave the lecture and left the room" and 'I left the room and gave the lecture" might be expected to have different meanings in normal conversation. In logic generally, as should be expected from the timeless nature of propositions, no such implication is intended, or expressible. The truth table for disjunction does no more than make its inclusive nature plain, but gives rise to some problems in that it allows disjunctions whose disjuncts which would not normally be acceptable. Thus "the White House is in Washington or New York" is true, given the truth table, because the first disjunct is true. We would not, however, say this under normal circumstances because the second disjunct adds nothing to our statement. The old joke of answering "Yes" to the question "Do you want tea or coffee?" illustrates a similar point. The respondent is using the *or* logically, the questioner in its more usual sense. Most dissimilar of all is material implication, since there is often assumed in ordinary language a connection, usually causal, between the two propositions. But in logic only the truth value is considered, so that 'Napoleon is the Emperor of China → the moon is made of green cheese" is true in virtue of the falsity of the antecedent, and "Napoleon is the Emperor of China → 2 + 2 = 4" is true in virtue of its consequent. Many surprising consequences of propositional logic turn on the attribution of natural readings to the operators, and the practice is to be deprecated.

Given the above rules expressed in the truth tables, we can construct a truth table for any expression in the propositional calculus, by evaluating each of the component expressions in turn. For example the truth table for (P → Q) ∨ (R ↔ − P) will be as follows, with the value of the whole expression given under its main connective, ∨:

P	Q	R	(P → Q)	∨	(− R	↔	− P)
T	T	T	T	T	F	T	F
T	T	F	T	T	T	F	F
T	F	T	F	T	F	T	F
T	F	F	F	F	T	F	F
F	T	T	T	T	F	F	T
F	T	F	T	T	T	T	T
F	F	T	T	T	F	F	T
F	F	F	T	T	F	T	T

From this truth table we can see that the complex proposition will sometimes be true and sometimes false depending on its constituents. Truth tables thus give us a means of determining the truth or falsity of an expression of arbitrary complexity.

3.2.3. Assignments and models

Each line of a truth table represents a different possibility for the truth value combinations of the constituent propositions. Each is thus an *assignment* of truth values to the propositions in the expression. In the case of the above expression the assignment {P: = T, Q: = F, R: = F} will make the expression false, whereas all other assignments will make the expression true. An assignment which makes the expression true is said to be a *model* for the expression; one which makes it false will be a *counter-example*.

Sometimes it may be the case that there is no counter-example to the expression, as in the following case:

P	Q	R	(P → Q)	∨	(R → P)
T	T	T	T	T	T
T	T	F	F	T	T
T	F	T	T	T	T
T	F	F	F	T	T
F	T	T	T	T	F
F	T	F	T	T	T
F	F	T	T	T	F
F	F	F	T	T	T

Such an expression will be true independent of the assignments made to its constituent propositional variables, and thus always true. Such an expression is called a *tautology*. Similarly, other expressions may have no model. These are termed *contradictions*. A simple relation exists between tautologies and contradictions: the negation of a tautology will be a contradiction, and vice versa.

3.2.4. Nature of a proof

A proof is intended to show that an argument is sound; this is so only if it cannot be the case that its premises are true and its conclusion false. Thus the argument Premises ⊢ Conclusion is sound if, and only if, Premises → Conclusion is a tautology. For the propositional calculus this can simply take the form of the production of a truth table, but other forms of proof can be given, using rules of inference. Since a truth table will contain 2^N lines given N propositional variables, the possibility of constructing a proof instead of using a truth table will be attractive in many cases.

3.2.5. Natural deduction

There are a number of rules of inference which allow us to prove arguments sound. Because they are attempts to formalise the way people reason intuitively they are called rules of natural deduction. Some of them will be mentioned here. First we may consider the rule which allows us to draw a conclusion from a conditional statement, given the truth of the antecedent, often called the law of *modus ponens*. Essentially this says that given a premise P → Q, and a second premise P, we can conclude Q. A second important rule is known as *modus tolens*. This enables us, given premises P → Q and − Q, to conclude − P. The two rules enable us to eliminate material implications from expressions, whilst others govern the introduction of material implication, the introduction and elimination of conjunction, and the introduction and elimination of disjunction, and the elimination of multiple negations (this last is simply that given $--P$, we may conclude that P). Also, importantly, there is the rule of *reductio ad absurdum*. This states that if we can prove a contradiction from one or more premises, then we have a proof of the negation of one of those premises. Thus given premises $P \to Q$ and $P \to -Q$, we can prove by *reductio ad absurdum* that $-P$, since if we assume the additional premise P, two applications of *modus ponens* and an introduction of & will give the contradiction $Q \& -Q$, which is a proof of the negation of our additional premise. It can be shown that these laws of natural deduction are *sound*, in that any conclusion derived using them will be true if the premises used are true, and *complete*, in that any consequence of a set of premises can be derived using them. Additionally it must be remembered that we can, as the possibility of using truth tables shows, always determine whether a given conclusion is, or is not, a consequence of a set of premises.

3.2.6. Equivalences between the logical connectives

Before leaving the subject of the propositional calculus, it is worth remarking that any of the expressions of propositional calculus can be written using only one of &, $\vee$ and $\to$, plus negation. This is because there are a number of equivalences between the various logical connectives. Thus,

$(P \to Q)$ is equivalent to $(-P \vee Q)$
$(P \& Q)$ is equivalent to $-(-P \vee -Q)$

and

$(P \vee Q)$ is equivalent to $-P \to Q$

These equivalences mean that we do not need all the connectives; we can re-write any expression in terms of $-$ and just one of $\to$, &, and $\vee$. Doing so, however, leads to more complicated and less natural expressions, and so natural deduction tends to prefer to use all the connectives. This possibility remains important, however, since normalising the expressions into some regular form is of great assistance in automated theorem proving, and so the existence of these equivalences is worth remembering.

3.2.7. Monotonicity

One other property of classical logic, common to both the propositional calculus and the predicate calculus, which will be introduced in the next section, is that it is *monotonic*. By monotonicity is meant that if a conclusion can be derived from a set of premises, that that conclusion may also be derived from any superset of those premises. This means, amongst other things, that no additional information can lead us to retract a conclusion. That classical logic has this property is easily seen: if we can construct a proof from a set of premises, then that proof will still be available given additional premises, since we can simply ignore them. If the additional premise contradicts one of the original premises, that does not matter, since classically we can derive anything at all from a contradiction. This property is important in AI since examples can be produced of standard reasoning techniques where it does not hold, where conclusions will be retracted on the basis of additional information. Such reasoning is said to be *non-monotonic*. This topic is further discussed in Section 10.3, which also describes some logics which lack this principle.

3.3. Basics of predicate calculus

The propositional calculus enables us to show sound a number of simple arguments, but unfortunately cannot even show the syllogisms recognised by Aristotle to be sound. For the soundness of many arguments does not rely only on the logical connections between propositions, but on relations between the constituent propositions, deriving, of course, not from their content but from their form. This in turn means that it is necessary to have a representation of propositions which allows us to see inside them and compare their logical forms.

3.3.1. Predicates

The most typical form of a proposition is the assertion that some property is true of some object; "John is fat", "the world is round", and the like. A natural way to represent the form of such propositions is to distinguish the two components, that denoting the property and that denoting the object. Such assertions are called *predications*, the property being termed a *predicate*.

3.3.2. Notation

This means that we must distinguish between the predicate and the object of which it is predicated. The convention used in predicate calculus is that the predicate is represented by an upper case letter (the sequence usually beginning at F), and the object by a lower case letter, the sequence usually beginning at a. The complete set of individuals which can appear in these expressions is called the *domain*. Thus, we can represent "John is fat" by Fa, the F representing the predicate "is fat" and the a "John". Now if we wish to represent three propositions such as "John is fat", "John is tall" and

"Tom is fat", we can do so by Fa, Ga and Fb; note that this shows up the common elements in the three expressions, whereas the propositional calculus would be forced to represent them as P, Q and R, concealing these common elements.

3.3.3. Propositional functions and quantifiers

This notation is adequate as long as we are making predicates about definite individuals. But we can see that F is on its own not well formed, but can be made into a proposition by supplying an individual. We can thus see the predicate as a propositional function, which when supplied with an individual as argument will become a proposition and thus map onto one of the truth values true and false.

Sometimes we will want to talk not about definite individuals, but about some class of individuals. Thus we may want to say that someone is fat, or all swans are white, or something similar. This is represented by completing the propositional function not with an individual, but with a variable; which is written as a lower case letter, the sequence usually beginning with x. Fx does not, however, make sense on its own, for we have no idea as to what terms representing individuals can properly be substituted for the variable. The expression is therefore completed by the addition of a *quantifier*, which may be the *universal quantifier*, written (∀x), indicating that any individual in the domain may be substituted for the variable, or the *existential quantifier*, written (∃x), indicating that one or more unspecified individuals in the domain may be substituted. Thus (∀x)Fx can be read as "everything is F", or "F is true of all individuals", and (∃x)Fx may be read as "there exists an individual which is F", or "something is F". We need to note two things about the quantifiers: first that if the domain is finite, we can express the universal quantifier as a conjunction and the existential quantifier as a disjunction. Thus if the domain is restricted to three individuals, a, b and c, then (∀x)Fx is equivalent to Fa & Fb & Fc; and (∃x)Fx is equivalent to Fa ∨ Fb ∨ Fc. Second, we should note that the quantifiers are closely related: (∀x)Fx is equivalent to − (∃x) − Fx and (∃x)Fx is equivalent to − (∀x) − Fx.

Quantifiers can also range over complex expressions; thus we can say that "all swans are white" with an expression such as

(∀x)(Sx → Wx)

and that "some swans are black" by

(∃x)(Sx & Bx)

Note that the connective in the existential quantification is & rather than →:

$$(\exists x)(Sx \rightarrow Bx)$$

would be true if there were nothing that satisfied Sx; whereas the sentence we were trying to represent implied that there was at least one swan. Sometimes we will want two variables in our expression, which will require two quantifiers: thus "all basketball players are tall and some cricketers are short" would be represented as

$$(\forall x)(\exists y)((Bx \rightarrow Tx) \& (Cy \rightarrow Sy))$$

This will be of more importance when we consider relations below.

3.3.4. Quantifiers and natural deduction

Normally the domain is taken as being infinite. The consequence of this is that we cannot use truth tables to decide the value of a complex expression of the predicate calculus. We can find models for an expression, but we cannot be sure that some other assignment would not lead the expression to evaluate to false, and we can provide counter-examples, but cannot be sure that some other assignment would not evaluate to true. There is no mechanical way of establishing the tautologous or inconsistent nature of the expression, and this is what is meant when predicate calculus is said to be *undecidable*. This gives an important role to deduction in the predicate calculus, since whilst we cannot show mechanically that an expression is tautologous we can often provide a proof which shows that it must be so. To do this we have to augment the natural deduction rules for propositional calculus by rules providing for the introduction and elimination of both the existential and universal quantifiers. The additional rules can be introduced so that the predicate calculus is both *sound* and *complete*.

3.3.5. Relations

In addition to propositions asserting that a property holds of some object in the domain, we also have propositions expressing relations between objects in the domain, such as "Tom is the father of John" and "the White House is in Washington". Now we could choose to represent these as simple predications; Fa, with F standing in the first case for "is the father of John", and in the second for "is in Washington". But to do this obscures some of the logical structure of the proposition; if we represented "Mary is the mother of John" by Gb, the connection with Fa representing Tom is the father of John would be lost. Therefore we introduce *relations*, written as upper case letters, normally in a sequence beginning with R. These are also propositional functions, but take more than one individual as arguments. Thus we can now write "Tom is the father of John" as Rab, and "Mary is the mother of

John" as Scb, revealing the presence of a common term in the two relations. Relations which take two arguments are called two-place or *binary* relations. The order of the arguments is significant, as each place in the relation denotes a different *role*, enabling the distinction between "Tom is the father of John" and "John is the father of Tom" to be expressed. Relations can take an arbitrary number of arguments, and the number of arguments is expressed as the *arity* of the relation. Thus a relation Rabcd would be of arity 4, and also sometimes called a *4-place relation*.

Relations can of course be used with quantifiers. Thus we may wish to express the fact that everybody has a father; this would be written as

$$(\forall x)(\exists y)Ryx$$

where R represents the relation "is the father of". The order of the two quantifiers is important: thus

$$(\exists y)(\forall x)Ryx$$

would express the quite different proposition that there is somebody who is the father of everybody (including himself!). This is an illustration of how the use of predicate calculus can eliminate ambiguity; whilst the above propositions have only one sensible reading, a proposition like "every student uses a computer" could mean either

$$(\forall x)(\exists y)Ryx$$

or

$$(\exists y)(\forall x)Ryx$$

according to whether every student used their own personal computer, or shared a common mainframe.

3.3.6. Identity

The above notation does not have a means of expressing the relation of identity of referent between two names standing for individuals. Thus whilst the use of different names is a necessary condition for denoting different individuals, we may discover that the names in fact refer to the same individual. Thus we may acquire a number of facts about Cicero and a number of facts about Tully, before we realise that Cicero and Tully are one and the same Roman. To meet this, predicate calculus is often extended to include identity, written as " = ", with its own rules for introduction and elimination to support proofs using it. The use of identity also allows us to express propositions such as "Everyone has exactly two parents", which we write

$$(\forall x)(\exists y)(\exists z)(Ryx \text{ \& } Rzx \text{ \& } -(y = z) \text{ \& } (\forall u)(Rux \rightarrow (u = y \lor u = z)))$$

This, admittedly rather clumsy expression, says that for any individual two distinct individuals stand in the relation R (parent of) to that individual, and that any other individual in the domain standing in that relation to that individual, must be identical with one of those two individuals.

3.3.7. Putting expressions into normal forms

For some purposes, especially when we consider automated theorem proving, the expressions permissible in predicate calculus are too diverse. For that reason a number of normal forms exist to make the syntax of expressions more uniform. It is important to note that any expression of predicate calculus can be transformed into these normal forms by a simple mechanical process.

The first of these normal forms is *conjunctive normal form* (CNF); expressions in CNF consist of a conjunction of elementary disjunctions. Thus it has the form

$$A1 \ \& \ A2 \ \& \ldots \& \ An \quad \text{with n greater than or equal to 1}$$

where each of the As is of the form

$$B1 \lor B2 \lor \ldots \lor Bb \quad \text{with n greater than or equal to 1}$$

Each of the Bs is either a proposition or its negation. Such propositions are often termed *literals*, with the proposition being a positive occurrence of the literal and negated propositions being negative occurrences of the literal. *Disjunctive normal form* is very similar, being a disjunction of elementary conjunctions. Thus it has the form

$$A1 \lor A2 \lor \ldots \lor An$$

where n is greater than or equal to 1, and each A has the form

$$B1 \ \& \ B2 \ \& \ldots \& \ Bn$$

where n is greater than or equal to 1, and each B is either a positive or negative literal. *Clausal form* consists of a conjunction of *clauses* where each clause has the form

$$A1 \lor A2 \lor \ldots \lor An \leftarrow B1 \ \& \ B2 \ \& \ldots \& \ Bm$$

with both n and m being greater than or equal to 0. The literals are always positive; clausal form makes no use of negated literals.

3.3.8. Method of transformation

Full descriptions of the method of transformation of arbitrary expressions into normal forms are given in Lemon [1], Jackson [3] and Clocksin and

Mellish [4]. Here I shall just work through a single example to give a flavour of the process. Suppose we wish to transform

$$P \rightarrow ((Q \rightarrow R) \rightarrow S)$$

into conjunctive normal form. First we use the equivalences that exist between the connectives to eliminate $\rightarrow$. (In particular here we use that $P \rightarrow Q \leftrightarrow -P \vee Q$.) This produces the expression:

$$-P \vee -(-(Q \vee R) \vee S)$$

Next we must drive the negations inside the brackets to get

$$-P \vee (--(-Q \vee R) \& -S)$$

and drop the double negations to get

$$-P \vee ((-Q \vee R) \& -S)$$

This is not in CNF because the & is subordinate to a $\vee$. We must therefore distribute the disjunction over the conjunction to get

$$(-P \vee -Q \vee R) \& (-P \vee -S)$$

which is in CNF. To convert this to clausal form we consider each clause in turn, and disjoin the negated literals on the LHS and conjoin the positive literals on the RHS, to get the two clauses:

$$R \leftarrow P \& Q$$

and

$$\Box \leftarrow P \& S$$

where $\Box$ represents absurdity. As another example of the movement from CNF to clausal form,

$$(P \vee Q \vee -R) \& (P \vee S)$$

would become

$$P \vee Q \leftarrow R$$

and

$$P \vee S \leftarrow$$

3.4. Exotic logics

The above all relates to what is called *classical* logic. But this is only one logic among many. Other people have, for a variety of reasons, developed different logics, either because they hold that they more faithfully reflect reasoning in certain domains, or to increase the expressiveness of logic. Such logics are often called *exotic logics* or *deviant logics*. There is no space here to do more than mention some: *intuitionistic logic*, held by some to be of particular relevance to mathematics, denies the equivalence of – – P and P. One of the consequences of this is that P ∨ – P is not tautologous in intuitionistic logic, although – – (P ∨ – P) is. This logic thus allows propositions to lack a truth value. Others have found two truth values insufficient and have wanted to introduce others to represent other notions supposedly lying between true and false. These logics are termed *polyvalent logics*. In some cases there is a fixed number of truth values, while others such as *fuzzy logic* allow any number of truth values; the truth value may be any number between 0 and 1.

A useful introduction to exotic logics is provided by Haack [5].

3.5. Modal logics

A rather different extension to classical logics are modal logics. These logics introduce operators which operate on propositions, perhaps treated classically. Thus for example we might use □P to represent something like "it is necessarily the case that P" and ◇P to represent "it is possibly the case that P" and develop a logic for these operators to reason about arguments involving these modalities. A good discussion of logics that can be developed for necessity and possibility is given in Hughes and Cresswell [6]. Other modal logics have been developed for other modal concepts: *temporal* logics dealing with, for example, "always" and "sometimes", and *deontic* logics, whose operators represent obligation and prohibition, being the most explored.

References

1. Lemon, E.J., *Beginning Logic*, Nelson, London, 1965.
2. Hodges, W., *Logic*, Penguin, London.
3. Jackson, P., *Introduction to Expert Systems*, Addison-Wesley, Reading, Mass., 1986.
4. Clocksin, W.F. and Mellish, C.S., *Programming in Prolog*, Springer-Verlag, Berlin, 1981.
5. Haack, S., *Philosophy of Logics*, Cambridge University Press, 1978.
6. Hughes, G.E., and Cresswell, M.J., *An Introduction to Modal Logic*, Methuen, London, 1968.

4
Search

The topic of search needs to be considered at this point for two reasons. Firstly because it has an historical importance in AI, in that the early AI programs were dominated by search issues. This means that many AI problems can be thought of in terms of search, and so a knowledge of search principles is an invaluable aid to understanding AI. In a sense, too, the increasing importance of knowledge representation is a response to the limitations encountered in the search approach to AI applications, so that an understanding of these limitations is helpful in motivating the study of knowledge representation. Even more importantly for our purposes, an understanding of how the knowledge representation paradigms we will consider later are manipulated will involve an understanding of basic search techniques. The object of this chapter is not to give an exhaustive account of search in AI, but simply to introduce enough of the terminology and techniques to enable an understanding of later material.

4.1. Search spaces

To get a feel for the basic search problem, it is helpful to consider the following not uncommon problem. Suppose we are in an unfamiliar town, without a map, and we wish to find a particular place in that town. All we can do is search for our destination by looking around. At each junction we will have a number of options: we may be able to go straight on, go back, turn right or turn left. At some junctions only some of the options will be available. Looking around will involve us in taking a series of these options until we reach our destination. If we are going to characterise this as a search problem we will need some terminology. Our presence at a junction is a *state*. Our starting point is the *initial state* and our destination is the *goal state*. Our route may thus be described as a succession of states, starting with the initial state and ending with the goal state, with the intervening stages being a succession of other states. We move from one state to another by selecting one of the options available to us in any given state, and that option will determine the successor state. Thus we may alternatively describe our route as a series of options which take us from one state to another. The options are termed *operators*, or *state transitions*. If more than one operator is applicable in a given state, we have to choose which operator to apply. This can therefore be called a *choice point*. If we decide we have made a mistake and retrace our steps to a previous choice point, we are said to *backtrack* to that choice point. The set of states which may be reached by applying legal operators to the initial state is known as the *search space*.

4.1.1. Characterisation of search problems

This terminology enables us to give a general characterisation of search problems. We are given an initial state and a goal state, together with a set of operators which, when applied to a state in the search space, will return a different state in the search space. The problem is therefore to find a sequence of states which will lead from the initial state to the goal state, and where each state can be reached from its predecessor by the valid application of an operator. This can also, perhaps more usefully, be expressed as a sequence of operators which, when applied to the initial state. will produce the goal state.

4.1.2. Examples

If we now apply this to the search problem of finding a particular destination in an unfamiliar town, we can see that the map as a whole represents the search space, and we want to find the ways to get from our initial location to our desired location. In this case the search space has a physical existence, in that it is the town, and the map provides a representation of the search space. Often, however, the search space will not have a physical existence, and we will not be provided with any form of representation of the search space at the outset. So we will need to generate the search space, or as much of it as we need to solve the problem, from the initial state and the operators. This process is well illustrated by the classic AI problem of the farmer, dog, chicken and grain.

In this problem we have a farmer returning from market, accompanied by his dog, having bought a sack of grain and a chicken. To reach his farm he needs to cross a river, and he must use a small boat to ferry himself, and his goods, across the river. The boat only has room for the farmer and one of the dog, sack and chicken. His problem is that if he leaves the dog and chicken on their own at any time the dog will eat the chicken, and if he leaves the chicken and the grain unattended, then the chicken will eat the grain. The problem is thus to find a sequence of trips across the river which will the get the farmer and his possessions to the other side, without any of them being eaten.

We can represent the occupants of the two sides of the bank as a pair of lists. Thus the initial state is ([F,B,C,D,G],[]). The valid operators all swap the "F" and the "B", and, optionally, one other term, from one member of the pair to another. The goal state is, of course, ([],[F,B,C,D,G]).

We can now generate the search space as shown in Fig. 4.1:

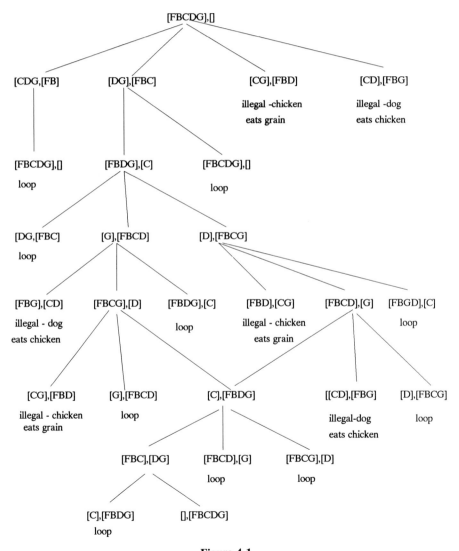

Figure 4.1

44 Knowledge Representation

We should notice here that the states we develop can fall into one of a number of categories. Sometimes a state which has already occurred at a higher level in the developing search space will appear again at a lower level. This represents a loop, and such a state need be developed no further, since we have got nowhere by the application of the intervening operators. Second, we sometimes reach a situation where the identical state appears on two branches at the same level. We need only develop one of these occurrences, since we should be indifferent to which path we used to reach the state because they are the same length. Third, a state may be illegal, in that it violates the constraints imposed by the problem. Such a state need be developed no further, since no such state can be on the route to an acceptable solution. Fourth, we may reach the goal state, in which case it need be developed no further since it means that the problem is solved. Finally, a state may fall in to none of the above categories, in which case we do need to develop it further, by the application of those operators which are valid in that state.

In the sort of very simple problem of the above sort we can find our solution simply by exhaustively developing the search space until we get to our goal state.

4.2. Search methods

More generally, however, we shall not be able to use this exhaustive technique, and even in some cases where we can it will be desirable to arrive at the goal state by means of generating as small a part of the search space as we can. To this end a number of different search techniques have been developed. What each of the techniques is really concerned with is how we go about generating the search space so that we may find the desired route from initial state to goal state without generating more of the search space than we need.

4.2.1. Goal-driven and data-driven search

One such thing we need to consider is whether it is really best to begin with the initial state, as we did in the above example. For each operator we can define an inverse operator which instead of taking us from a state to its successor would take us from a state to its predecessor. If we use the set of inverse operators, we can begin with the goal state, and find the path to the initial state by the successive application of these inverse operators. Consider the following example of a popular sort of children's puzzle (Fig. 4.2).

If the puzzle is framed in terms of what Jack caught, then we should start at the beginning and follow Jack's line forward until we come to his catch.

Figure 4.2

But if the problem is who caught the boot, it is better to work back from the goal state, the boot, and see to whom the line belongs, since this will mean that we need follow only one of the lines, whereas two times out of three we will guess wrongly if we start with the fishermen, and thus need to follow two or three lines.

Similarly, when generating our search space, we can either start from the initial state and apply operators looking for the goal, in which case we can say we are using *data-driven search*, or we can start from the goal and look for a suitable initial state, when we say we are using *goal-driven search*. Other names for these two kinds of search are *forwards reasoning* and *backwards reasoning* respectively. Which we use depends on our view of the nature of the search space. If we suspect that the branching around the initial state is greater than that around the goal state, we should use goal-driven search, whereas if the branching is greater around the goal we should use data-driven search.

This is illustrated by the following example taken from the *blocks world*. The *blocks world* is an example domain much used in AI, consisting of a number of blocks of different shape and colour on a table which need to be re-arranged to form a desired configuration. Suppose we have the situation

46   Knowledge Representation

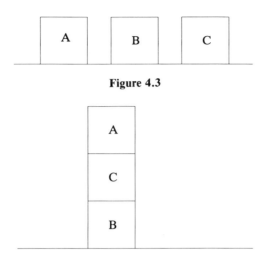

Figure 4.3

Figure 4.4

in Fig. 4.3 as our initial state and we wish to bring about the configuration in Fig. 4.4. Our legal operators all move a block to a new location, either another block or the table, and can be applied only if the block to be moved is clear. Let us represent states by a list of three letters, representing the positions of blocks A, B and C respectively. If the block is on another block, this will be represented by the appropriate letter, and if it is on the table, this will be represented by a T. Thus the initial state will be [TTT] and the goal state will be [CTB]. If we start working forwards from the initial state we generate the search space in Fig. 4.5. However, if we start from the goal state and work backwards, we get the search space in Fig. 4.6. The latter is greatly to be preferred, because in that case there is initially only one legal move, whereas from the initial state there are six. Data-driven search thus involves a proliferation of nodes representing poor initial moves.

4.2.2. Breadth-first, depth-first, and bounded depth-first searches

As well as the direction in which we search, we need to consider the order in which we develop the nodes we encounter in the search space. We might think that we have to be systematic about this, and, if so, there are two obvious principles which can be used, and which give rise to searches with different characteristics. We might choose to expand fully nodes in the order in which they are encountered, in which case we will perform a *breadth-first search*, or we could develop the most recently encountered node, in which case we will perform a *depth-first search*.

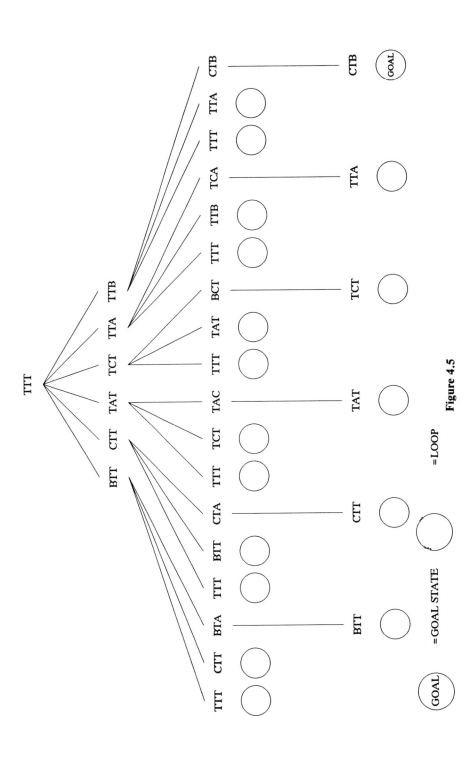

Figure 4.5

48   Knowledge Representation

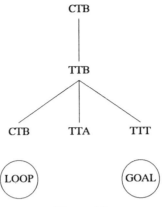

Figure 4.6

Depth-first search is the sort of search that corresponds to what we do when we look around an unfamiliar town. Because in such cases we develop a node by physically moving to that node, we have no choice but to continue our search by developing that node next. Sometimes this will involve us in reaching a dead end or loop, in which case we will have to backtrack and retrace our steps to the most recent choice point and continue searching by developing that node in some different way. Developing the search space using a depth-first method produces the situation in Fig. 4.7, in which the nodes are numbered in the order of their occurrence.

With depth-first search we only ever have one node open at a time (the node we are at in the looking-around case), and this has the advantage of being economical on memory. There are, however, two problems with depth-first search. First we may set off on entirely the wrong track. That is, the path through the search space that we explore first may not contain the goal state at all. If the path terminates, this is not too serious, since we shall eventually have to backtrack, and so get on a path which does contain a solution. It may be, however, that the path does not terminate, in which case we will not find a solution at all, but follow the wrong path for ever, or at least until the resources of our machine run out. It is therefore sensible to modify our procedure so as to limit the depth to which we will explore a given path, and backtrack when that bound is reached, regardless of whether we have failed on that path or not. This modification is known as *bounded depth-first search*. The problem here is that the solution will be found only if it exists within the depth bound, since otherwise we should be forced to backtrack before we arrived at the solution on that path. So the price we pay for ensuring that a path terminates is that we may not go far enough down it. This flaw can be remedied if we modify the procedure yet again so that once

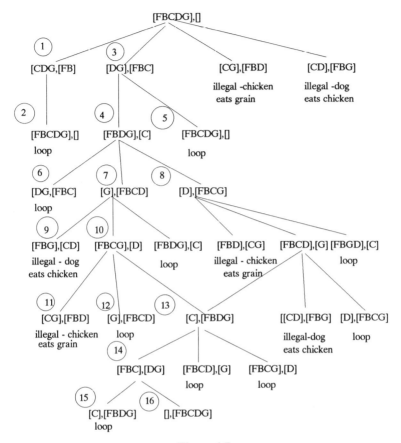

Figure 4.7

we have explored the entire search space to the given depth we do the search again with an increased depth bound if no solution has been found. If we do this we must choose between two further evils; either we store all the leaf nodes of the unsuccessful search, which is costly in terms of memory, or we start again from scratch, in which case the original exploration must be performed again, which is costly in terms of time. The moral is that, if we use bounded depth-first search, we need to choose a depth bound which is sufficiently high to make this need to restart unlikely.

The other problem with bounded depth-first search is that whilst it will find a solution, if one exists within its depth bound, we can have no assurance that the solution found is optimal, in the sense of being the solution with the shortest path. For, it may be that a solution exists on several paths, and had we chosen a different path it might have proved shorter, as in

50 Knowledge Representation

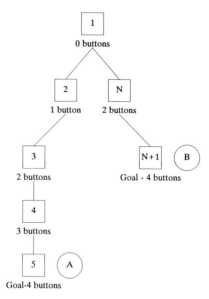

Figure 4.8 *Problem*: collect 4 buttons. *Actions*: pick up 1 button or pick up 2 buttons. Depth-first search finds solution at A but misses better solution at B.

the situation in Fig. 4.8. Of course, we could explore these paths as well, looking for better solutions, but then some of the advantages of depth-first search are lost. For this reason it is preferable to use depth-first search only when there is no reason to insist on the shortest-path solution.

The alternative method of breadth-first search is illustrated in Fig. 4.9.

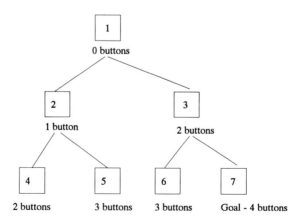

Figure 4.9 Breadth-first search for button problem of Fig. 4.8.

Here we explore the search space to a given depth before expanding any nodes to a greater depth. The consequence of this is that we avoid the problems associated with depth-first search, because we are assured of finding a solution if one exists, and because we are assured also that there can be no shorter path to a solution, because all shorter paths have already been explored. Thus breadth-first search might seem preferable, except that we need to be aware of just how many nodes might need to be open and available in memory as we explore the search space. For any problem of realistic complexity, it is certain that we will exceed the memory resources of our machine. This is an important reason why, in practice, depth-first is the method that is normally used.

4.2.3. Heuristic or ordered search

As can be seen above, neither of the above systematic search methods seems to supply the ideal answer. In realistic problems, the search space is very large, and neither method can cope with this complexity. Take chess playing as an example: white has 20 options available to him on his first move, as does black. Thus after a single turn by each player we have 400 positions to consider. The number of options available on the second move depends on the first move made, but they tend to increase (particularly if the first move is a good one) rather than decrease. When we additionally consider that a solution (a mating position) is rarely reached in fewer than 30 to 40 moves, given reasonable play on both sides, we can see that brute force alone is unlikely to allow us to progress very far.

Two reactions are possible to this; either we can abandon the whole enterprise, or we can try to find a way of doing rather better than systematic search. After all, a chess player does not proceed by exhaustive and systematic examination of the search space; rather he can detect that certain moves are promising and others are bad simply from a consideration of the position that will result from making them, and he will, of course, only give further consideration to positions which look promising. The lesson is that, given that the developing search space will have a number of nodes available for expansion, we should expand that node which seems to us, at the time, to be the most promising, and we can also simply discard very unpromising nodes. This is the idea that underlies *heuristic search*, a heuristic being the means by which we estimate which will be the best node to expand.

As well as ordering the nodes we wish to develop, we have the possibility of ordering the operators we will apply to a given state. Thus a chess player may have as a maxim that the knight should always be moved when it is possible to do so, and so commence his search by looking at possible knight moves. Thus a heuristic search can be ordered on state evaluation, where nodes are

52 Knowledge Representation

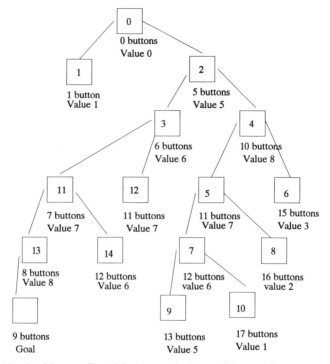

Figure 4.10 *Problem*: collect 9 buttons. *Actions*: pick up 1 button or pick up 5 buttons. *Evaluation function*: 9 – absolute value of (9 – buttons held).

evaluated, or on operator evaluation, where the possible operators are evaluated, or on some combination of the two. In practice we find the most common approach is to evaluate the states, and then to apply all possible operators to the chosen node to generate the new set of states to consider.

The search space for a heuristic search with state evaluation may look something like Fig. 4.10. Notice how the inappropriate evaluation function leads to the exploration of an incorrect path. A node initially highly rated becomes less highly rated when expanded.

4.2.4. The 8-puzzle as an example of heuristic search

One of the classic AI problems that heuristic search has been applied to is the so-called 8-puzzle (Fig. 4.11). This is a scaled-down version of the common children's toy, the 15-puzzle, which consists of 15 sliding tiles in a 4 × 4 grid. The idea is, given a random starting position, to re-arrange the tiles by sliding them into the vacant position so as to attain a sorted configuration such as shown in Fig. 4.12.

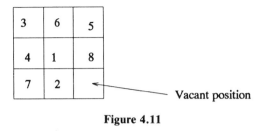

Figure 4.11

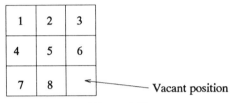

Figure 4.12

Even the 8-puzzle is too complicated to solve by brute force methods, and so it makes a good vehicle for experimenting with heuristic search ideas. The key problem is to develop an *evaluation function* which will enable us to assess the promise of the nodes in the developing search space. There are a number of evaluation functions that we might apply to the various states of the 8-puzzle. A very crude one would be simply to count the number of misplaced tiles. This can, however, quickly be seen to be unlikely to be very successful, since we could improve our score simply by moving a tile to its correct location, but that this would, in fact, take us away from the desired solution if it meant that it was blocking the path of another misplaced tile, because we would have to move it again in order to solve the puzzle (Fig. 4.13).

A second method of evaluation would be to count and sum the distances of each tile from their proper squares. This would, of course, fall foul to the same objection as the first suggestion, but it would at least recognise that not

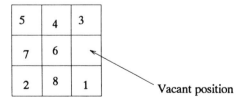

Figure 4.13 Moving the 6 is pointless because either the 6 or the 8 would have to move to allow the 1 to move.

all good moves resulted in a tile being put in its proper place. Yet another refinement, this time bringing us closer to a good solution, would be to add to the sum resulting from the last suggestion the number of tiles not followed by their proper successors. This factor takes into account the difficulty of moving the tiles about.

All of these evaluation functions would allow us to assign some score to a node. But this is not by itself enough to choose a node for expansion. For suppose that we had two nodes with equal scores on the evaluation function, but that one was reachable by a shorter path than the other. If the paths from the nodes under consideration to the goal are of equal length, as they would be given an accurate evaluation function, the total paths will be of different lengths. Therefore we should expand the node which can be reached by the shorter path: and we need to score nodes by adding the cost of reaching the node to the score returned by the evaluation function.

Of course, we could attach quite arbitrary numbers to our nodes, but the above consideration suggests a natural interpretation for the return value of the evaluation function. If we let the evaluation function return our estimate of the number of steps required to move from the node being evaluated to the goal node, then this combined with the number of steps taken to reach the current node will represent the length of the path from initial state to goal state which goes through the node under consideration. Note that this means that we will prefer the lower scores.

So popular did this idea prove that a considerable amount of mathematics was done exploring the resulting search algorithm. In particular there was some interest in whether the algorithm would find the shortest path to a solution. It can be proved that if our evaluation function represents a lower bound on the actual number of steps to the goal, then the solution we find using the algorithm, and such an evaluation function will be optimal. It is unnecessary to go into the proof here [1].

Whilst this is a nice mathematical result, finding such an evaluation function is difficult in practice, and may not lead to an optimal search if we are more interested in minimising the number of nodes examined in the finding of the solution than in the number of nodes on the eventual path found. This can be seen by considering that an evaluation function which returns zero in all cases will represent a lower bound, and will find the optimal solution, but it will generate a simple breadth-first search, and so buy us nothing.

In practice, therefore, it was often better to lose this nice property in order to get a more efficient search. In the example of the 8-puzzle, a good evaluation function proved to be the sum of the distances from the home square plus 6 times the number of tiles not followed by their successors, plus 3 if there was a tile in the centre. This did not represent a lower bound on the steps from node to goal, and so the solutions it found were not provably optimal. None the less it proved to work better, in terms of finding solutions, that other evaluation functions.

The moral of this is that although we do, of course, want to find the best solution, and to expand the fewest nodes needed to find that solution, in practice, we will settle for finding a solution whilst expanding a practicable number of nodes. The best strategy to adopt was invariably bound up with the nature of the problem—the size of the search space, the branching factors the likely depth needed to find a solution, and the availability of an effective evaluation function.

4.3. Limitations of search

As the subject of AI developed, it began to be recognised that search alone was not enough. There were several factors which made it unsatisfactory as a general problem-solving technique.

4.3.1. Search space size

First there is the sheer size of the potential search space for problems that can be considered interesting. Note that the example above was the 8-puzzle rather than the 15-puzzle we commonly give our children to play with. And then imagine applying similar techniques to try to build a program that would solve Rubik's cube. For many—probably most—problems we are tempted to solve, the search space is too big to enable even an efficient search program to solve them in acceptable time.

Note that I am talking about problem-solving programs here: chess programs still do rely heavily on search. In this kind of game-playing problem the object is not to find a solution, but to find an acceptable move. Therefore we simply look ahead as far as time allows and return the best move we find. This is not necessarily on the path to a solution (a forced mate), but is good enough to give most human chess players a hard time.

4.3.2. Evaluation function problems

There were also problems with heuristic search, which although it did mitigate part of the size problem which rendered brute force search out of the question for substantive problems, put great emphasis on finding a suitable evaluation function.

Not only were such evaluation functions difficult to find, and hard to establish other than by trial and error, but they were invariably entirely specific to the problem being addressed, so that lessons learnt (often

56 Knowledge Representation

painfully) on one area were not readily transferable to new problems. In some cases, such as chess, this was unimportant since the problem mattered enough to justify the effort devoted to solving it. In other areas, however, the objection is fatal; the 8-problem is of little interest if not a stepping stone to the 15-puzzle.

Second, the whole approach fell into difficulties since in some cases the use of the evaluation function failed to lead to a solution, but rather got diverted into the pursuit of purely local optima. In a large class of problems there is no smooth progression from the initial state to the goal state. See Fig. 4.14 for an example. In such cases, there would be no effective way of using the search method to find a solution.

A final objection, relevant if we are interested in AI as a computational model of the mind, is that such methods do not model, or even attempt to model, human reasoning when confronted with problems of the sort we are considering. Except in the simplest class of case a human is unlikely to systematically search the domain, and certainly no human would use the sort of evaluation function thrown up by the 8-puzzle work to solve such a

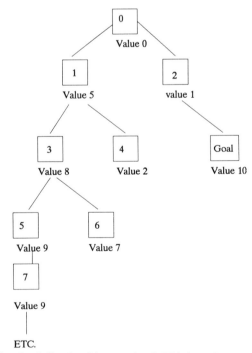

Figure 4.14 All nodes following 7 have value 9. This is a plateau, which means that no solution will be found nor will backtracking occur.

puzzle. Thus such solutions, even when they worked, were running counter to some of the important motivations underlying AI. They were essentially using non-intelligent methods to solve the problem, and hence were attempting to reduce what had seemed problems requiring intelligence, to problems which could be solved mechanically.

For this reason interest in search in its pure form declined, and attempts were made to see whether a study of how humans go about solving these problems might provide some insights which could be incorporated in AI programs.

4.4. Human problem solving

This study of human problem-solving methods gave rise to the observation that three types of problem-solving knowledge are used by the human problem solver. I shall make these distinctions, because I think they are there to be drawn, and because I think it gives a useful framework for considering problem solvers. None the less, much of the work in AI would not draw these distinctions, and instead uses the term "heuristic" to cover all three sorts of problem-solving knowledge.

4.4.1. Decomposition of problems

The first is a tendency, when confronted with a problem of reasonable complexity, to attempt to decompose it into a sequence of subproblems. These subproblems may be simple enough to tackle straight away, or they may themselves be decomposed into subproblems of their own.

Thus, if you present a person with a Rubik's cube for the first time, an instinctive way of solving it is to attempt to get one face correct first, then a second, and so on. This can be used to illustrate an important point: there will usually be a number of different ways of decomposing the problem, and some will be far more effective than others. In the cube example, getting a face correct is not usually a good step; a better technique is to get the corners correct and then the other pieces. One of the reasons why Rubik's cube is a hard puzzle is because, whilst the instinct to decompose the problem is there, seemingly natural decompositions turn out to be counterproductive.

If we now apply this to the 8-puzzle, we could decompose this into position the 1, position the 2 and so on. In fact, a better decomposition is first to get the first row right, then the first column, and then shuffle the final three tiles around until a solution is found. Once found this knowledge can be transferred to other forms of the puzzle. Thus a good decomposition for the 15-puzzle is get the first row right, get the second row right, get the first column right, get the second column right, and then shuffle the last three pieces around. Or to generalise further: given an N by M grid, get the first N-2 rows right, get the first M-2 columns right and then shuffle the last three pieces.

Getting the first row right is not, however, entirely straightforward, and itself requires some sensible decomposition. This might be to get 3 following 2 and 2 following 1 and then move the resulting chain into place. Again this can be generalised.

It is instructive to compare the way the problem can be decomposed with the final evaluation function given above. Clearly, following the strategy outlined here will tend to improve the result from the evaluation function, both by getting tiles to their home squares and by increasing the number of tiles followed by their successors. But it will not necessarily, in any given case, effect the greatest improvement. The human problem solver is not really concerned to find an optimal solution, but is more concerned that he is at all times clear about what he is doing, and can see a steady flow of achievement as he completes subtasks which he knows are steps on the road to a complete solution. It is this kind of factor, involving a strategic rather than a purely tactical appreciation, which avoids the kind of problems that we encountered in Section 4.3.2.

Knowing how to decompose the task into the right subtasks gives us a strategy for solving the problem, but is unlikely to take us all the way to a solution. For we still need to know how to operate at the tactical level, so that we can achieve the subtasks. This is where a second feature of human problem solving comes in.

4.4.2. Search for patterns

As well as breaking the task into subtasks, problem solvers also try to identify patterns with which they are familiar, and to which they know how to respond. In the 8-puzzle example we can consider a subset of the grid, as in Fig. 4.15. It is quite often the case that we can have a fragment of the board like Fig. 4.15(a) which we want to re-arrange so that it looks like Fig. 4.15(b). We also know a sequence of moves which will effect this re-arrangement, namely the sequence: move Y, any, X,Y, any, X,Y, any. This sequence can be applied wherever in the grid the fragment occurs, and whatever the size of the whole grid. The skilled solver will have knowledge of many such patterns and appropriate move sequences, and he solves the problem by decomposing the problem into subtasks where such patterns can be identified and appropriate sequences performed. Incidentally this is also the way to go about solving Rubik's cube, where the whole trick of mastering the cube is to know a series of moves which will exchange two particular pieces, or re-orientate a piece, whilst leaving the other parts of the cube unchanged.

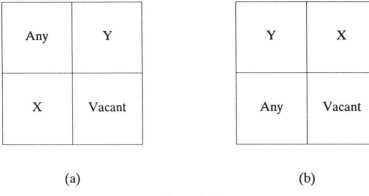

Figure 4.15

This two-part solution method is quite general: decomposition supplies the strategic knowledge of how to go about solving the problem, reducing it to a set of tasks which can be performed by recognising patterns to which the appropriate actions can be taken.

4.4.3. Rules of thumb

There is a third type of human problem-solving characteristic which can be observed, which might be termed the use of "rules of thumb". This knowledge represents actions which are usually, but not always, good things to do in certain circumstances. They are thus very like the patterns mentioned above, but need to be applied more opportunistically rather than as the execution of a set design, and, crucially, they are recognised to be fallible. A rule of thumb in the 8-puzzle would be that if a tile is in the central position, it would generally be a good idea to move it. The justification for the rule of thumb is that it increases the number of moves available, and we should notice that it also improves the evaluation function used above. Within the framework we have been developing, rules of thumb should only be applied as an effort to reach a familiar pattern where none can be seen.

If we wish, we can introduce a general notion, a *heuristic*, to cover all three of these kinds of problem-solving knowledge. We can see a heuristic as a matter of recognising a situation and making a certain response to that situation. The situation may be simple and the response uncertain in effect, as in the case of what I termed "rules of thumb" above; or the situation, and the response to it, may be more complicated and the effect of the response more certain, as in the case of what I termed "patterns". Or the situation may be a task to perform, and the response to perform a series of subtasks, as in what I called "problem decomposition". This way of seeing things suggests that we can represent all three types of problem-solving knowledge in the same way, and the reader will recall that uniformity was a desirable feature of a knowledge representation.

In particular we can represent all three types of knowledge in the following way using a uniform mode of expression:

> Rules of thumb such as "If tile in the centre, then generally move that tile"
> Patterns such as "If pattern A obtains and goal is goal 2, then perform move sequence 6"

Note the use of the qualifier on the action in the case of the rule of thumb whereas the pattern is thought to be always applicable. Finally we can represent decomposition as

> "If goal 1 then achieve subgoal A and subgoal B"

An attempt to follow through this method of knowledge representation provides us with the first major knowledge representation paradigm that we shall examine, namely *production rules*.

Exercises

4.1. A robot designed to collect specific numbers of buttons can perform three actions. It can

(a) pick up 1 button;
(b) pick up 5 buttons;
(c) drop 1 button.

It is given goals of the form "collect n buttons" and uses a sequence of its actions to perform its task. It is a precondition of action (c) that the robot holds at least one button.

   (i) Draw the complete search space for the task when the task is to collect 9 buttons, assuming that search is breadth-first and that actions are selected in the order (a), (b), (c). Explain what is meant by depth-first search, breadth-first search and bounded depth-first search.
   (ii) For each of the three search strategies named in (i), say under what conditions they will find a solution, supposing that at least one exists, and whether the solution found will be optimal. Explain your answers.
   (iii) Would breadth first or depth first search be more appropriate for the above problem? If bounded depth-first search was used, what would be an appropriate depth bound?

4.2. (i) For the above problem devise an evaluation function which could be used to perform a heuristic search. Does it find a shortest-path solution?
   (ii) Find a suitable evaluation function for the problem in Fig. 4.8. Does it differ from that used in (i)? If it does, explain why.

4.3. You are studying a map of Liverpool on which are marked 12 public houses, 4 restaurants and the Everyman theatre. You wish to work out a route from where you are to a public house (you don't mind which), from the chosen public house to a restaurant (you don't mind which), and the Everyman. Say for each stage whether data-driven or goal-driven search would be more appropriate.

Reference

1. For an excellent discussion of this and other matters relating to search, see Nilsson, Nils J., *Problem-Solving Methods in Artificial Intelligence*, McGraw-Hill, New York, 1971.

5
Production Rules

At the end of the last chapter we saw how the various sorts of knowledge used in solving puzzles such as the 15-puzzle could be expressed in the form of if-then rules, associating some pattern expressed as a set of conditions and goals, with some actions which will help to realise those goals. This insight led to the development of an important knowledge representation paradigm which has become widely used in artificial intelligence, namely *production rules*. In this chapter we shall examine this paradigm in detail.

5.1. Form of production rules

Production rules have two components, which are often represented as two lists. The first component is a list of one or more conditions and the second component a list of one or more actions which may be appropriately performed if and when the conditions are satisfied. Informally they may be read as "IF conditions are satisfied, THEN perform actions". Because they have these two parts, conditions and actions, production rules are sometimes called *condition-action pairs*.

5.1.1. Entity-attribute-value triples

The form in which the various conditions and actions are expressed may vary, but there is one common form which merits specific consideration. A condition usually involves some entity having a certain value for some attribute of that entity, such as John being 18 years of age, the 4 tile being three moves away from its home square, and so on. When we come to use the production rules in a system, we shall need to determine whether the conditions are satisfied. Such an investigation will be facilitated by uniformity in the expression of the conditions. For this reason the conditions are often expressed as lists of three elements: the entity involved, the attribute of the entity under consideration, and the value of that attribute for that entity. Such lists may be called *entity-attribute-value triples*. So we could represent the two examples above as "(john age 18)" and "(4-tile distanceFromHome Square 3)".

This format is really surprisingly flexible, since it is in effect a notation for binary relations, named by the attribute name, and holding between an entity and a value. In order to express complex relations we must be able to express binary relations, since unary relations would be expressively inadequate. If, however, we can express binary relations, our representation is expressively adequate to deal with relations between any number of entities since we can re-write relations of any arity in terms of a set of binary relations. Consider a relation of arity 3, giving, as in "John gave the apple to Mary". Using a relational notation this would be written as something like "gave(john, apple1, mary)". If we wish to re-write in terms of binary relations we must introduce another entity, the giving event, and use this entity to relate the other three entities which participated in the event. Thus we can express the above relation as the following three binary relations.

> (giving1 donor john)
> (giving1 gift apple1)
> (giving1 recipient mary)

The new entity, the giving event, has three attributes which take the participants in the event as values. This technique, introducing a linking entity and writing a separate binary relation, named for the role in the relation, for each of the entities in the original relation, will work for relations of any arity. If we wish to, we can also re-write unary relations, such as "John is fat", as binary relations either by using the copula as a relation name as in "(john is fat)", or by using some boolean value as the value, as in "(john fat true)".

Thus we can render any set of relations uniformly as sets of binary relations, at the cost of some proliferation of entities and relations. The uniformity will prove useful when we come to use the relations in a computer system, because we will be able to process all our relations in the same way without any need to consider different arities, and this may well be worth the cost in terms of the reduction in brevity and clarity. The important thing to recognise is that adopting this restriction leads to no loss of expressive power.

5.1.2. Variables in production rules

This representation is fine for expressing conditions which refer to facts about specific entities. But this is rather restricting. In the 8-puzzle we may want a rule which says that if a tile is in the centre position we should move it. We could say

(1-tile position centre), (move 1-tile)
(2-tile position centre), (move 2-tile)
and so on for all eight tiles

This is not only tedious but it also fails to capture the knowledge we are trying to represent. For the rules, considered individually, would suggest that there is some peculiarity about the tile mentioned in the rule which means that it should be moved if it is in the centre position. Only when we consider the rules as a whole do we see that the principle holds good for all the tiles. But that the principle is true of any and all tiles is the very knowledge that we wished to express. What this means is that we need our representation to allow us to talk about classes of things as well as individual things. This in turn means that we require the notion of a variable which can stand impartially for any one of a group of things. Variables can be written as a letter preceded by a question mark. Thus the above set of rules can be rewritten using a variable as the single rule

(?x position centre), (move ?x)

When we come to use such a rule the variable will be replaced by a particular entity. Notice that the variable occurs twice within the rule; when we substitute an entity for the variable it is important that such substitutions are made uniformly, so that the variable is replaced by the same entity for all its occurrences in the rule. We can describe this process of substitution either by saying that the variable is *instantiated* to the entity substituted for it, or that the variable is *bound* to the entity.

In the example the variable is used to stand for an entity. This is the most usual, and the most useful, case, but there is no reason in principle why variable should not stand for the other elements of our entity–attribute–value triples, should we find it useful to do so. Thus if we wished to express the fact that brothers have the same parents we could write the rule:

IF   (?x has-brother ?y), (?x has-parent ?z)
THEN (?y has-parent ?z)

When the conditions of the rule are satisfied, the variables in the condition part of the rule will have been replaced by the appropriate entities and values, and so we will know, from the need to substitute consistently, that any parent of ?x is also the parent of any brothers that ?x may have.

We could also, if we wished, allow variables to stand for attributes. This is, however, of less likely utility, and sensible examples are hard to come by. Perhaps we could express the knowledge required to imitate someone as

IF (?x imitating ?y) and (?y ?z ?w) THEN (?x ?z ?w).

If we do this we should be aware that we are going beyond what can be expressed in first-order predicate calculus, since this effective quantification over attributes is a feature of the second-order predicate calculus. We need to be both very careful about what we are actually saying when we do this, and aware that we are very likely to find the processing of such rules inefficient. In practice variables are usually permitted only for entities and values.

5.2. Components of a production system

We must now turn from looking at production rules purely as a notation for representing knowledge and start to consider how they might be used to solve problems within a computer system. In outline, what is required is a system which will take a set of production rules, test whether the condition parts of the rules are satisfied, and execute any appropriate actions, and continue with this cycle until the problem is solved. Such a system will in general have three major components, each of which will need to be discussed. These components are the *working memory*, the *production memory* and the *rule interpreter*.

5.2.1. Working memory

In order to apply the rules that we have developed to a given situation it will be necessary to have a notion as to what the situation is. That is, we will have to know what the facts of the situation are, and what the goals we wish to achieve in that situation are. The function of working memory is to act as a "scratch pad" on which we can record these things. Thus to begin with the working memory will record the facts which describe the initial state, and the goals that we wish to satisfy from that state. Thus, if we are using triples, working memory will comprise a collection of triples describing the current facts and a collection of triples describing the goals, the facts we wish to realise. As we proceed towards a solution, the facts will change, which may necessitate the addition or deletion of the facts in working memory, and we may decide upon additional goals, which must be added to working memory, and achieve goals which in consequence will be goals no more and so can be removed from working memory. Thus the contents in working memory are essentially dynamic; they represent the momentary situation, and so as the situation changes so items will be added to and deleted from the working memory. This feature is particularly useful for representing actions which alter the facts of a situation; a robot for example will have a location which will change as it performs a task. Only the current location needs to be considered, and the over-writing of the location attribute in working memory will provide a natural and efficient way of representing these changes.

This leads to an important point. When we think of actions we often think of real actions, such as the movement of a robot arm, or the repositioning of some object. These are, of course, legitimate actions for the action parts of production rules. But we should not forget that in order for the results of actions to become known to the system is is necessary to update the working memory to record the effects of these actions. Therefore the action part of the rule must always contain actions which do this updating of working memory. So if we have an action which repositions some object, then the action part of that rule must also delete from working memory the current position of the object and add to working memory the resulting position of the object.

In many systems moreover, there may be no "real" actions performed at all, and the reasoning will take place solely in terms of the working memory of the system. In such a case all the actions in the action part of the rule will be actions which do no more than update the working memory. In such a case the representation of the actions can be simply in terms of the triples which should be added to or deleted from the working memory. Rules written in this style will have less of the appearance of condition–action pairs, and more the appearance of logical implications, but the principle remains the same provided that one recognises that the actions are internal to the system. Here the reasoning style is very like ordinary deduction. It should, however, be noted that the dynamic nature of working memory is less well fitted to this quasi-logical reasoning. Strictly speaking, deductive consequences, in a classical logic framework, are cumulative and do not cease to be true. This means that we need to be very wary of deleting information from working memory, as this behaviour goes against this feature of classical logic. This possibility of non-monotonic behaviour in production systems is a very important issue, which has both positive and negative ramifications, and which will be discussed in detail later in Section 10.3.

5.2.2. Production memory

The role of the production memory is to record the production rules that we wish to bring to bear to solve the particular problem on hand as represented in the working memory. In contrast to the dynamic nature of the working memory, the production memory is essentially static. These rules will be of general applicability, and therefore can remain constant, both throughout the consideration of a particular problem, and from problem to problem.

Note that this general applicability does not require that the rules must not contain references to particular entities. We may have a rule of the form

> IF (john dinnerGuest true) THEN, (?dinnerMenu vegetarian true)

to record the fact that a particular person does not eat meat. This rule, although it contains no variables in its condition part, is of quite general applicability, since it will be appropriate for any dinner menu-devising situation where John may be a dinner guest.

5.2.3. Rule interpreter

We now have the situation where we have all the facts and goals relevant to the problem in hand recorded in working memory, and the rules that we can bring to bear on the problem in production memory. The final component of the system is therefore the one which will apply the rules to the given situation so as to perform the actions determined as appropriate and make any consequent updates to the working memory. This process of applying the rules will continue until the problem is solved.

What then must the interpreter do? It must go through each of the rules in its production memory in turn, attempting to match the triples in the condition parts of the rules with triples in working memory. If all the triples in the condition part of a rule do so match, then the actions in the action part of the rule are candidates for performance. It may be that a number of rules from production memory have their condition parts satisfied in this way, and in this case it will be necessary to make a choice as to which action part is to be performed. The set of rules which are candidates for execution is called the *conflict set*, and the process of selecting a rule to execute is known as *conflict resolution*. Executing the action part of the rule is called *firing* the rule (perhaps an extension of the analogy by which the satisfaction of the condition part is said to "trigger" the rule). Once a rule has been chosen and its action part executed, there can be a check to see whether all the goals in working memory have been satisfied. If they have, then execution can cease, otherwise the processing of matching and firing will need to be repeated. This cycle of matching and firing is sometimes called the *recognise–act cycle*.

5.3. Operation of a production system

Thus far we have conceived of our interpreter as selecting the rules to appear in the conflict set by finding those rules whose condition parts are composed of triples that are facts in our working memory. The effect of this strategy is to produce data-driven search, since which rules will fire will depend on the facts of the initial situation, and will have no reference to the goals that we have with regard to the problem we are trying to solve. As we saw in the previous chapter, however, many problems are more suited to search which is driven by the goals. We can adapt our interpreter so that it will operate in this way.

5.3.1. Goal-driven search using production rules

To produce goal-driven search we generate our conflict set not by matching the condition parts of rules with the facts in working memory but instead by matching the action parts of the rules with the goals in working memory. If a goal corresponds to a fact which can be added to working memory as an action of a rule, the rule tells us that were its condition parts to be satisfied, then this action would be performed, the requisite fact added and the goals satisfied. Thus we can add the triples in the condition part of the rule to our goals in working memory, since we now know that when they have been achieved, the original goal will have been achieved. Thus we can delete the original goal (and any other goals in working memory which also appear in the action part of the same rule) from the working memory. In effect the goals have been replaced by the subgoals required to achieve them. In addition a goal need not be added to working memory if it already appears as a fact in working memory, since it can be taken as already achieved.

An example may help to make things clearer. Suppose we have the following set of rules in production memory:

a   If T1 and T2 then T3
b   If T3 and T4 then T5
c   If T3 and T5 then T6

and the facts T1, T2 and T4 as facts in our working memory. Suppose that our goal is to achieve T6. We therefore add T6 to our goals in working memory. Rule c will enable us to achieve this goal, and so we can add T3 and T5 to our goals, and delete T6. Rule b will allow us to achieve T5, so we can make the goals in our working memory T3 and T4. But T4 is already known as a fact, so we have no need to add it to working memory. Our goals are therefore just T3. Rule a will enable us to achieve T3, so making our goals T1 and T2. But these are already facts in our working memory so we need not add them, and our goal list is empty, and we can stop, having achieved all our goals.

In this mode of operation, the rules give the appearance of being used backwards, in that we are using desired actions to generate more goals from the condition parts of the rules that would produce them. Moreover, as new goals are generated, this gives rise to new rules that can help us solve the problem. For these reasons goal-driven search in rule-based systems is often described as *backward chaining*, as a chain of rules from goal to initial state is generated.

5.3.2. Data-driven use of production rules

If we reconsider for a moment the more obvious application of production rules, namely the data-driven search which arises from matching the condition parts of the rules, we can see here that as we add facts into working memory we enable more rules to match. Thus the firing of one rule will cause other rules to match, generating a chain of rules forwards from the initial state to the goal. For this reason, data-driven search in a rule-based system is often termed *forward chaining*.

5.3.3. An example

At this point we can look at a very simple example. Suppose we are recording some knowledge about where we should go to buy various common articles. We go to the newsagents to buy newspapers, crisps and matches, and to a grocers to buy cornflakes, crisps and milk. Of course, we need to also know that the shop is open, and that newsagents are open every day, but that grocers are only open on weekdays. We can represent the knowledge as a set of production rules thus:

```
EPR1 IF (?x wants ?y)
     AND (?z sells ?y)
     AND (?z open true)
     THEN (?x go-to ?z)

EPR2 IF (?x is-a newsagents)
     THEN (?x sells papers)
     AND (?x sells crisps)
     AND (?x sells matches)

EPR3 IF (?x is-a grocers)
     THEN (?x sells milk)
     AND (?x sells crisps)
     AND (?x sells cornflakes)

EPR4 IF (day is weekday)
     AND (?x is-a grocers)
     THEN (?x open true)

EPR5 IF (day is ?x)
     AND (?y is-a newsagents)
     THEN (?y open true)
```

Now we can suppose that there are some facts in our working memory, such as (Roberts is-a grocers), (Heaths is-a grocers), (Billings is-a newsagents), (day is sunday) and (John wants crisps). If we now run the rule interpreter in a data-driven manner on the above, several of the rules will match and more facts about what is sold at the various establishments will be added to working memory, as will the fact that Billings is open and that John should go to Billings. In this case we have drawn a large number of unnecessary conclusions, such as that Roberts sells cornflakes, and we might consider it better to run the system in a goal-driven manner. If we add the goal (John go-to ?x), EPR1 would match adding goals (John wants ?y), (?z sells ?y) and (?z open true). (John wants ?y) is satisfied by a fact in working memory with ?y bound to crisps. Now EPR2 and EPR3 would match; rules would continue to fire, and eventually the ?x in our original goal would become bound to Billings.

5.3.4. Control and conflict resolution

In the discussion above about the rule interpreter we have seen that the interpreter generates a set of rules which could be applicable in the situation as represented by working memory. Sometimes there will only be one rule which is applicable, and then we will have no problem, because we will have no choice: we simply fire that rule. If no rules are applicable we do have a serious problem (assuming we have not yet attained our goal), because we cannot proceed. In the general case, however, we will have a set of rules—the conflict set—and we are forced to choose which one we are going to apply. Thus in the example above, when working in a goal-driven manner, we had to choose between firing EPR2 and firing EPR3 on the second cycle.

When we remember that rules will modify the working memory, we will see that this choice is quite significant since the application of one of the rules from the conflict set will mean that on the next cycle there may be new rules which are applicable. If this was all there would only be efficiency considerations to worry about, since eventually we would be able to try all the rules in the conflict set. But, since we can also delete facts from working memory, rules may also cease to be applicable, because their triggering conditions have been deleted, and so we may have only one chance to fire a given rule. Thus the choice of the rule to fire not only affects the efficiency of the problem solver, but may be responsible for the problem being soluble or not, in cases where a bad selection of rule to fire results in a rule we needed being no longer applicable. The problem of deciding on a strategy for choosing a rule from the conflict set to fire is thus of crucial importance in the construction of a production-rule system, since it has an impact on the operation of the system both in terms of its problem-solving behaviour, and its problem-solving capacity. The topic of selecting a rule from the conflict set is known as *conflict resolution*.

5.3.5. Conflict resolution strategies

In choosing our conflict resolution strategy we will be guided by a number of concerns. First we will wish to avoid loops. Without any conflict resolution strategy it is easy for loops to develop. For if a rule does not vitiate at least one of its conditions it will reappear in the next cycle. If it was chosen the first time, there is no reason to suppose that it will not be chosen again. If this is so, we will endlessly perform the same actions over and over again. In the example above, EPR1 always forms part of the conflict set given the goal (John go-to ?x), but continually applying this rule will not advance us to a solution. Thus avoiding loops is a minimum requirement of a conflict resolution strategy.

This can be done by insisting that a rule does not fire twice on the same set of data. (We cannot simply ignore a rule once it has fired as we may wish to use it with different bindings for variables, or because its conclusions may get deleted from working memory by some other rule, and then it may be required again.) This principle is sometimes called *refractoriness*. This can be implemented either by discarding a rule which has fired once from the conflict set altogether, or, as a weaker form, in discarding the rule which fired on the last cycle from the conflict set.

Other conflict resolution principles are concerned with promoting convenient behaviour in the system. There are two major concerns; first we wish to be responsive to changes in the situation and second we wish to follow one line of thought until it proves to be the wrong line of thought. Both of these principles can be implemented by preferring rules in the conflict set generated by items most recently added to working memory. Thus if the application of a rule added an item to working memory, then rules whose conditions parts contain that item will be preferred in the next cycle, whereas rules applicable in virtue of previously added data will wait. This will obviously make the system responsive to changes in the working memory, but it will also have the effect of following a train of thought. The general effect is to produce a depth-first search. For the purpose of firing a rule is to take a step towards a solution, and if it was a good choice, in that it is in fact a step towards the solution, then its conclusion will be required for the solution, and the next step will require that conclusion as a condition. Of course, it may well be necessary to backtrack and fire the rules grounded in older data if this path proves not to lead to a solution. The behaviour of a system using this principle will therefore be similar to one which uses depth-first search as its guiding principle. This principle is often called *recency*.

To illustrate the use of recency, let us consider the example again. Here the first conflict set is simply EPR1 and no conflict resolution is needed. The next conflict set will contain all the rules since all match with a goal in working memory. Recency will reject EPR1, since the goals (?z sells crisps) and (?z open true) will have been added to working memory since (John go-to ?x). If recency is the only principle, we will need to make an arbitrary choice between them. Suppose we choose EPR3: this will add the goal (?z is-a grocers), satisfied with ?z bound to Roberts or Heaths, and so adding goals (Roberts open true) and (Heaths open true). Again all our rules will appear in the conflict set, but recency will ensure that we attempt to establish whether Roberts and Heaths are open before firing EPR2 and considering the newsagents. Thus we are sticking with a single line of enquiry before following other paths, effectively generating the search space in a depth-first manner. We would, of course, have found the solution more quickly had we chosen EPR2 in preference to EPR3. Recency, by itself, is not enough to ensure that we have a determinate answer as to which rule to fire, and so it needs to be augmented by some additional principle, if we are to remain in control. One very simple answer is to choose the rule first encountered in the production memory, and this may enable to programmer to increase efficiency by ensuring that the more promising rules occur higher in production memory.

A third popular principle of conflict resolution is one known as *specificity*. This essentially states that we should prefer rules which are harder to satisfy in that their condition parts contain more conditions, or fewer unbound variables. Thus if one rule in the conflict set contains four conditions and another rule only three, we could use this principle to select for firing the rule with four conditions. The intuitive basis for this is that the rule with the greater number of conditions is more finely tuned to the situation under consideration, and can be seen to its best effect when the conditions of the rule with the fewer conditions are a subset of the other rule. The presence of this conflict resolution procedure is often exploited by people designing production-rule knowledge bases. Consider for example the following two production rules:

1. IF (?x bird true) THEN (?x canFly true)

and

2. IF (?x bird true) AND (?x is-a penguin) THEN (?x canFly false)

Rule 1 is a convenient falsehood: convenient because most birds can fly and because we do not wish to enumerate all the flying birds, and a falsehood because there are exceptions to this rule. If we have specificity as a conflict resolution principle, no harm will be done by including rules such as 1 which are strictly speaking untrue, provided we cover all the exceptions with rules such as rule 2. The extra conditions in the rules describing the exceptions will ensure, given this conflict resolution principle, that the inapplicable general rule never fires if the conditions in the rules covering the exceptions are satisfied, since the exceptions will always be associated with a greater number of conditions.

5.3.6. Defeasible rules

This introduces us to a point of great importance. A good deal of the knowledge that we may wish to incorporate in our knowledge-based system may be "rules of thumb" of the kind represented by rule 1; namely good general principles which are not, however, infallible. If all we know about a creature is that it is a bird, then it is natural to conclude that it can fly. We therefore need a way of representing these defeasible principles in any general knowledge representation schema.

The use of specificity does, as explained above, enable this from the point of view of expressiveness. There is however, a price to pay in that if we choose to do this we have rules which express something which is false unless we have the appropriate other rules in the knowledge base. Thus rule 1 is only expressing a truth, even on the assumption that specificity will be used as a conflict resolution principle, if all exceptional cases are covered. We might feel confident in our ability to provide a rule for each type of flightless bird, but this would be not enough, since a normally flying bird might in fact be flightless owing to some exceptional circumstance. To use a well-known example, Eddie might well be an eagle, but he still will not fly if his feet have been set in concrete.

We could, of course, simply ignore this last failing, and say that we have no desire for our knowledge base to handle such unlikely cases. We can do this, and our program will continue to give the right answers, provided none of the uncovered exceptional cases arise. None the less a purist could argue that we have not represented the knowledge pertinent to the situation, since we effectively change the meaning of the default case whenever we modify the knowledge base by adding another exceptional case that we feel it is necessary to handle. This lack of independence in the rules moves us away from the ideal of a declarative representation of knowledge, and it must be remembered that it was on this ideal that many of the advantages claimed for this sort of system were based.

There will be more discussion of this point in Section 10.3 when we come to deal with non-monotonic reasoning: all we need to note for now is that the operation of conflict resolution is such that we cannot know the effect of production rules in a knowledge base without knowing both the conflict resolution principle that is going to be employed and the other rules in the knowledge base. It is these two things taken together with the apparent meaning of the rule, rather than a declarative reading of the rule considered in isolation, which will determine how our system behaves.

To return to the topic of conflict resolution: we have seen several principles that could be used, and all of them have some merits, but, with the exception of an appeal to the order of the rules in production memory, none can be relied on to select a single rule. Practical systems tend to use a combination of all these principles. The production rule language OPS5, for example, uses the following algorithm for conflict resolution. First it discards instantiations (conditions with variable bindings substituted in them) that have already fired. Next it compares the recencies of the working memory facts that match the conditions. If some instantiations are tied, it then compares the number of conditions in the instantiations. If there is still a tie it chooses the first encountered. OPS5, however, does offer a second strategy which is a refinement of this, in that after discarding used instantiations, it first considers the recency of the *first* condition in the instantiations and only proceeds to consider the other components for those which tie on this criterion. By giving this extra importance to the first condition it increases efficiency, since fewer comparisons of recency need be made, and gives the programmer additional control, in that he can exploit this in his choice of order for the conditions in the rules. Of course, there may be an impact on the behaviour of the system, and so this extra control does mean an additional loss of declarative purity. Readers who wish to find out more details of conflict resolution in OPS5 should consult Ref. 1.

5.4. Pros and cons of production systems

The major advantage of the production-rule paradigm is that it appears to provide a single uniform method of representation, which form of representation is relatively easy to understand for the non-computer specialists that we expect to find our experts drawn from. The IF . . . THEN format has a meaning which is intuitively straightforward, and the use of triples (particularly if presented in a more English-like way by providing some linking text between the items) is not difficult to grasp. A second claimed advantage is that such a paradigm facilitates incremental development; that is, refinements, and additions, to the knowledge represented in the knowledge base can be easily incorporated by refinements, or additions, to the existing rules. Furthermore, the rule interpreter is relatively straightforward to implement, and can be made quite efficient if one incorporates some easily available methods such as the Rete matching algorithm [2], which computes the additions and deletions from the conflict set during each cycle rather than computing it from scratch each time.

These advantages have been claimed, but each of them needs to considered with some scepticism. Whilst the meaning of individual rules can be grasped by non-computer specialist experts, such people do tend to have difficulty in going beyond the declarative meaning of the rule, and considering the implications that flow from the presence of other (perhaps conflicting) rules in the knowledge base, and the conflict resolution principle that will be employed by the rule interpreter. The inability to weigh the truth of rules in isolation is a distinct disadvantage as far as ready comprehensibility goes.

With regard to incremental development, whilst it seems that we can just go on adding rules, once we realise that these rules are not truly declarative we do need to worry about what other rules there are, and how, given our conflict resolution strategy, they will interact with the new one. Large production systems do tend to become unwieldy and hard to update, so that the lack of structure starts to become a drawback.

Finally, the simple matching and selection procedure employed by production systems can give rise to unsatisfactory behaviour which needs to be tuned by giving the programmer more control over the order of firing. The use of contexts in XCON, discussed in Section 9.5.2, and the general issue of the representation meta-level control knowledge discussed in Section 10.6, will illustrate this point.

Exercises

5.1. Consider question 4.1 of the exercise for Chapter 4.
   Write a set of production rules to enable the robot to pick up exactly n buttons in as few operations as possible. Say whether the rules should be run in a forward or backward chaining manner, and whether any conflict resolution strategy is used. Indicate which rules will fire and what will happen to the contents of working memory when the robot is asked to collect 9 buttons.

5.2. You are writing a production-rule system to do some elementary medical diagnosis. The doctor tells you that a person with a runny nose has a cold, unless his eyes are swollen, when he has hay fever, except if he has blue spots on his face as well, in which case he has Vinny's syndrome.
   Write a set of production rules to capture this expertise, and suggest a conflict resolution strategy which will give the appropriate behaviour.
   The doctor supplies the additional information that if the patient has red spots on his face, he has measles, unless his eyes are swollen, in which case he has Rose's disease, except if his nose is running too, when he must have Harvey's Complication. Expand your system to include this additional information. How valid are the claims of easy incremental development of production systems?

5.3. (i) Describe (briefly) the important characteristics of a "production system".
   (ii) Discuss the issue of *control* within a production system.

5.4. (i) Write a short paragraph to explain what is meant by the following terms as used in production rule expert systems:
   (a) production rule;
   (b) working memory;

(c) backward chaining;
(d) forward chaining;
(e) conflict resolution strategy (mention two possible strategies).
(ii) A machine has, among other components, three levers and two lights. The lights are controlled by the levers according to the following table:

Lever 1	Lever 2	Lever 3	Red light	Green light
Up	Up	Up	On	On
Up	Up	Down	Off	Off
Up	Down	Up	Off	Off
Up	Down	Down	Off	Off
Down	Up	Up	Off	On
Down	Up	Down	Off	Off
Down	Down	Up	Off	Off
Down	Down	Down	Off	Off

In addition a bell rings whenever both lights are on.

Write the production rules for this part of the system. Make clear any conflict resolution principle the rules rely on. State whether any situations are not covered by the rules.

5.5 (i) What is meant by
(a) data-driven search;
(b) goal-driven search.
(ii) What factors need to be considered when choosing between data-driven and goal-driven search? How do these considerations apply to MYCIN and XCON
(iii) What is a production rule? Explain how each of the above search strategies is implemented when knowledge is represented in the form of production rules.

References

1. Brownston. L., Farrell, R., Kant, E. and Martin, N., *Programming Expert Systems in OPS5*, Addison Wesley, Reading, Mass., 1985.
2. For a description of the algorithm see Forgy, C.L., Rete: A fast algorithm for the many pattern/many object pattern match problem, *Artificial Intelligence*, **19** 17-37 1982.

6
Structured Objects

So far we have been mainly concerned with the rule-based representations of knowledge, but now we need to consider also some popular alternatives. In this chapter, therefore, we shall look at two varieties of what can be collectively referred to as structured object representations (following Nielsson [1] and Jackson [2]). The two types we shall consider are *semantic networks* and *frames*. Both of these representations have their roots in efforts to build systems which understand natural language, such as simple stories, and are motivated by a desire to correspond in some way to human understanding. The starting point is therefore somewhat different from that of the rule-orientated representations which derive in the main from problem-solving applications. This difference in origin, however, does not alter the fact that the goal is the same, namely the representation of knowledge required for the task. Further, both representations are to be found in contemporary systems, as are combinations of the structured object and rule-based paradigms.

6.1. Semantic networks

Semantics is a central part of the study of linguistics, along with syntax and pragmatics. Whilst syntax says what combinations of symbols in the languages are allowed, semantics is the study of the meanings of those symbols, and pragmatics relates the use of utterances of the symbols to the intentions of those making the utterance. The study of semantics in particular, therefore, is an attempt to describe the meanings of words, and semantic networks attempt to give this description by relating the symbols in a network. A network is a graph (in the mathematical sense), which comprises a number of nodes linked together by arcs, with both nodes and arcs having labels. In a *semantic network* (usually abbreviated to *semantic net*), the nodes are supposed to represent concepts denoted by the words in the language under consideration, and the arcs relationships between these concepts. A simple example of a semantic net is given in Fig. 6.1.

80 Knowledge Representation

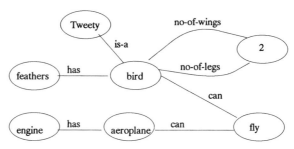

Figure 6.1

In the net of Fig. 6.1 there are captured both a definition of the concept of a bird, expressed by the arcs saying that a bird is a biped and a bird has feathers, and associations between concepts in that it expresses that both birds and aeroplanes can fly. Also, the net can be used to record facts about the world; the use of the proper name "Tweety" as a node label enables us to express the fact that Tweety is a bird. The semantic net therefore gives us a static representation of some knowledge of the world, and of the concepts used to describe that knowledge. The ideas behind this representation have a fairly lengthy philosophical history which may, some claim, be traced back as far as Aristotle, who claimed both that human behaviour is controlled by learnt associations between concepts, and that concepts are built out of more primitive concepts. This tradition has extended down to the present day through the British empiricist philosophers such as Locke and Hume and their contemporary followers. The use of these ideas in AI, however, can be effectively traced back to work done by Ross Quillian in the 1960s [3]. Those interested in early developments of semantic net ideas should consult this reference together with Winston [4] and Schank [5].

6.1.1. Origins

As mentioned above, semantic nets have their origin in systems motivated by the desire to understand natural language, as opposed to problem-solving systems. Thus while the predominant sort of knowledge motivating production rules was that expressing relationships between situations and what actions might be taken in those situations in order to achieve some desired goals, the concern motivating work on semantic nets was recording of the relationships between objects described in the sentences of the story under consideration. Much of the information regarding these relationships was noted to be implicit in the sentence. Thus, for example, if I say that I gave Mary a bicycle there is information implicit in this statement to the effect that I also gave Mary all the components normally taken to constitute a

Structured Objects 81

bicycle, so that I will also have given Mary two wheels and a saddle. This means in turn that someone who understood my original statement should be able to respond correctly to a question as to how many wheels did the vehicle I gave Mary have, and see a question as to what the saddle was made of as being pertinent. This is, of course, a very different sort of knowledge from that concerning appropriate inferences and actions which may be made or taken from a description of a state of affairs to solve a given problem. It is therefore unsurprising that semantic nets should have a rather different feel from production rules.

6.1.2. Features

To get a better understanding of semantic nets and the ways in which they can be used, consider the following example. Suppose we wish to represent the information in some simple story such as "John (7) gave Mary (6) a blue book. She sold it to Jane (7) for $10." We might do this by creating a net like Fig. 6.2.

We can see that we have a number of nodes representing the principal actors in the story and their attributes and a number of arcs relating these nodes. The diagrammatic representation is attractive to many who find it easier to understand the various relations pictured in this way than when they are expressed written in a purely textual form, like the entity-attribute-value triples we saw in the last chapter. If, however, we consider what is being expressed by the diagram, we see that they have a simple correspondence to these triples. The arcs of the semantic net correspond to the attributes in the triples, and the nodes in the semantic net correspond to the entities and attributes of the triples. Thus the semantic net, like the triples, represents a

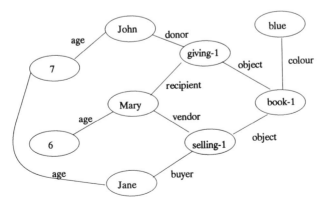

Figure 6.2

set of binary relations, and the expressive power is equivalent. One difference worth noting, however, is that in the net the nodes representing the entities and values have the same status; thus while in the triples the value may or may not be something which also occurs in entity position, in the semantic net all values have the status of entities.

Another important difference lies in the treatment of general terms. In the triple notation, we were allowed to use variables to stand for general terms, and to make use of these to express information true generally of all members of a class. Thus if we wanted to say that all bicycles have two wheels, we would do so by relating the appropriate triples through a production rule with variables such as

IF(?x is-a bicycle) THEN (?x no-of-wheels 2)

In the semantic net we have neither variables nor the rule formalism. Instead we have nodes representing general terms, and the links from these nodes express this generally applicable information. Thus a fragment of a net representing, amongst other things, bicycles, might look like Fig. 6.3. We therefore represent both facts and generally applicable truths in the same formalism, and the need for rules stating information applicable to all members of a class diminishes.

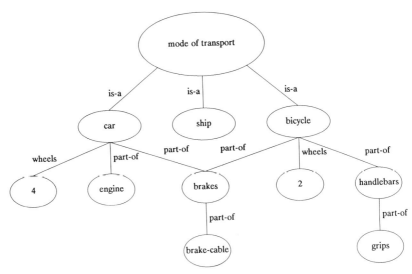

Figure 6.3

6.1.3 Inference in semantic nets

We have thus far suggested that semantic nets and production rules are expressively similar in that both represent a set of binary relations about individuals and general terms. We need now to see how a system can manipulate the representation to derive new conclusions from it.

Returning to the example of a bicycle, let us suppose John gives Mary a blue bicycle and we want to know how many wheels it has. In a production system we would have the following:

> IF (?x is-a bicycle) THEN (?x no-of-wheels 2)
> (giving-1 donor John)
> (giving-1 recipient Mary)
> (giving-1 gift bicycle-1)
> (bicycle-1 is-a bicycle)
> (giving-1 is-a giving)
> (bicycle-1 colour blue)

If we now pose the question (bicycle-1 no-of-wheels ?y), the variable in the rule will become instantiated to bicycle-1, and an application of the rule will instantiate the variable in the query to 2. Now let us see how the process works in a semantic net. The situation is described by the net of Fig. 6.4. Now we can follow the is-a link from bicycle-1 to the general term bicycle, and the no-of-wheels link from bicycle to the value 2. The process is much the same, although the processes of instantiation and rule application have been

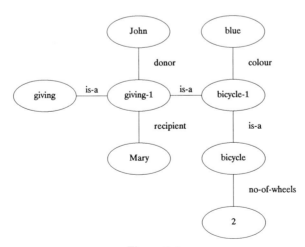

Figure 6.4

replaced by following the appropriate links. To give a second illustration, suppose we want to know the colour of John's gift to Mary in Fig. 6.2. We would, in a production system, need to pose a query something like:

(?x is-a giving) AND (?x donor John) AND (?x recipient Mary) AND (?x gift ?y) AND (?y colour ?z).

Note that in this case there is no inference as such needed; the complex nature of the query is required to ensure we pick out the right act of giving and so instantiate the variable ?y and hence the variable ?z to the correct values. Using a semantic net we would again proceed by following links. We would need to follow the links from the giving node until we found one which had a donor link to John and a recipient link to Mary, and to follow the gift link to bicycle-1 and the colour link from this node to find the required answer. All this was usually expressed in some procedural code (typically LISP) written to manipulate the semantic net in question, and accessed through some appropriate front end provided by the system.

And this brings us to an important point. Although the usual way of presenting semantic nets is in the form of diagrams which people may use their navigational skills to interpret, within a computer this will be represented (assuming LISP has been used) as a set of atoms for the nodes, which have been associated with pointers to other atoms to represent the links, and the inference will be made by certain pointer-following procedures. These procedures can be implemented quite efficiently in LISP using property lists, which was one of the attractions of semantic nets to early AI workers. A problem, however, is that different links may need to be followed in different ways; thus "is-a" is transitive, so we can happily follow is-a links through any number of intermediate nodes. A relation like "near" is not, however, transitive: whilst Portsmouth is near Southampton and Southampton is near Bournemouth, Portsmouth is not near Bournemouth. Thus we cannot follow links representing the "near" relation in the same way as links representing the "is-a" relation, and will need to use different procedures in the two cases. If we allow many link types, and one of the attractions of semantic nets is the opportunity it gives to do this, we will need to write many different procedures to follow them. The inference mechanism within semantic nets is thus more diverse and complicated than is the case with production rules, where a uniform inference mechanism can be employed, the diversity of inferences appropriate to different sorts of relations being catered for by the fact that permitted inferences are expressed in the rules. Hence, while semantic nets may seem to gain by not requiring a large set of rules, in fact the need is no less present, but is hidden in the inference mechanism. The apparent simplicity is thus only apparent, and in fact the opacity of the inference mechanism may lead to problems. For example, the relation "caused-by" is only arguably transitive: over a very long chain of causes it becomes implausible to suggest that the origin of that chain was the cause of the final effect. In a semantic net, the relation will have a precise interpretation in this respect according to how the relevant procedure has been implemented, but this is not readily available to the user of the system, whereas in a production rule system the likely inferences could be seen from an examination of the rules concerned. This is of particular significance when the names used for the various links may be susceptible to several, subtly different, interpretations. For example, in the semantic nets drawn above, is-a was used to relate a particular bicycle to the general term bicycle. If the net had been more complicated, there may have been a node "mode of transport", and it would have been tempting to use is-a to link the general bicycle node to this node. This, however, would have been a mistake since the relation between two general terms and a general term and an instance of it are quite different in a number of ways. For an excellent discussion of these points and others related to them, see Woods [6] and Brachman [7].

Therefore, if we wish to get all the expressive power from semantic nets that we can get from a rule-based system, we will need this extensive collection of pointer-following procedures, perhaps even one for each link type in the system, which seems to be an argument against them. It should, however be said in defence of semantic nets, that this is not always necessary for the uses to which they are put. In Quillian's original system, for example, the types of queries posed generally required no more than the determining of the relation between two nodes. This can be done by a simple and efficient procedure, known as *intersection search*. Here the two nodes in question are activated and this activation spread along the links from those nodes until a node in the net is activated from both directions. This will indicate the shortest path between the two original nodes, and hence the association between them.

There is then a trade-off between the flexibility of the system, the specificity of the questions which we can ask, and the complexity of the underlying manipulation mechanism, and the consequent effects of the intelligibility of the representation.

6.1.4. Psychological justification

Another feature which some have seen as an advantage of semantic nets is that the representation tends to cluster information relating to an object around that object. This not only has useful implications for computational effectiveness, but is also claimed to be the way human memory works. One of the original motivations for semantic nets was to answer the question of how meanings of words could be stored so that human-like use of these meanings could be made. Quillian developed an associational model of memory [8], and it is this model that his semantic nets reflect. Whether or not one accepts this model as a model of human memory, however, there can be no doubt that one may find the notation, particularly its diagrammatic form, very attractive. Indeed, when trying to get an understanding of a domain many people find themselves drawing diagrams which look very like semantic nets. This, however, is not a persuasive argument for using them inside an implemented system: there different considerations apply. In particular the reason why they are easy to use to sketch out an area is just that they enable someone to proceed without needing to think too precisely about the exact nature of the links which he is using. As an aid to understanding this is legitimate, but when encoded in a system the operation of the system will supply a precision which may not have been intended. A successful system cannot be implemented without addressing these questions of detail, therefore. Thus while progress may be initially faster, difficult questions may still have to be answered.

6.2. Frames

Frames (a term popularised by Marvin Minsky [9]) represent another form of structured object which can be seen as a development from semantic nets. In semantic nets there is a certain lack of structure, which leads to realistically sized nets becoming extremely complicated. There is therefore a need to impose some kind of structure on the representation to make it tractable. One of the sources of complication is that the nodes themselves are all alike (apart from their labels) and that there are very many of them. Also, if one looks at nodes with a given label one can see that they typically have the same set of arcs emerging from them. Thus any giving event will have arcs labelled donor, recipient and gift (and perhaps location and time as well). The basic idea of frames is to differentiate various types of node according to the relationships they typically participate in, and to hide this information within the node, so that it becomes visible only when we wish to see it. Take the example of John giving the bicycle to Mary in Fig. 6.4: we could represent it in the following way, using three frames:

```
giving-1
  is-a: giving
  donor: John
  recipient: Mary
  gift: bicycle-1
bicycle-1
  is-a: bicycle
  colour: blue
```

bicycle
　　is-a: mode of transport
　　no-of-wheels: 2

What we have done here is to list the nodes with more than one link, and to group the link names and the values of them under it. This reflects one of the intuitions motivating semantic nets, namely that information should be clustered around the objects, but it simplifies matters by pushing many of the relationships inside the node. It also gives the possibility of handling the general term/instance distinction in a better way. When we describe a situation we tend to refer to instances rather than general terms. If John gives Mary a bicycle, we know that it is an instance of the general term bicycle, even if we know no more about it. Thus we know in turn that it can participate in all the relationships which bicycles participate in, even if we do not know with which things it participates in these relationships. Specifically, we know, for example, that it has some colour, even if we are ignorant as to which colour it happens to have. This opens the opportunity to avoid the confusions that can arise within semantic nets of treating general and specific terms as if they were on the same level. Instead we can regard general terms as things which are instantiated by particular objects. In the example above, colour is proper to the instance bicycle-1, but this fails to reflect the knowledge that all bicycles have some colour. Two-wheeledness is proper to bicycles, but this is proper to the general term bicycle, rather than a property which happens to be true of all bicycles. Representation using frames gives us the possibility of reflecting these intuitions: that all members of a class have certain attributes whether we know the value of them or not; that for some attributes all members of the class will have particular values for certain relationships, and that there is a critical distinction between instances of general terms and the general terms themselves.

6.2.1. Basic ideas

Following on from that brief informal introduction to frames, we can say that the basic ideas underlying a frame representation are as follows. First there is the notion that frames correspond to objects. These objects will participate in a number of relations represented by *slots* within the frames. In turn the objects participating in these relationships will be filled by values; which may themselves point to frames, or to some simpler structure. In the example above, John, Mary, bicycle-1, bicycle and mode-of-transport are all intended to be frames, 2 to be an integer and "blue" to be a string. To compare with entity–attribute–value triples: entities correspond to frames, attributes to slots and values to the contents of the slots, often called *fillers*.

Frame-based representations, because of the central role of slots and fillers, are sometimes called *slot and filler* representations.

In the last paragraph I said that both bicycle and bicycle-1 were intended to be frames. In fact there is a distinction, made by most frame systems, to be drawn here, since bicycle is the general term and bicycle-1 the particular term. Bicycle can be viewed not as a frame in the strict sense, but rather a schema describing any frames that are instances of this general term. Indeed *schema* is a term used by some in preference to frame. The schema, representing the general term, will determine the slot possessed by its instances, and in some cases (as in no-of-wheels above) will determine the values of the slots in those instances, whereas in others (such as colour above), the value will be assigned only at the instance level. Making this distinction requires us to reflect it in our representation, and in the example above it might be better signalled by the use of "a-kind-of" instead of "is-a" as the slot name in the bicycle frame.

Thus knowledge representation using frames involves the identification of the types of object that will be of interest to the system, and describing these classes of objects in terms of frames/schemata. These schemata will contain information as to what attributes objects of that class have, by defining a set of slots, and, where the values of those attributes are determined by class membership, the values for those slots. When the system is run, and information about particular individuals becomes known, these schemata will be instantiated to particular frames with the appropriate slots. As information about these individuals becomes known, those slots not filled in virtue of class membership will be assigned appropriate values. Retrieving information will be a matter of getting the values from appropriate slots, and may involve a certain degree of indirection through a number of frames before the desired value is reached. Thus if we wish to access the colour of John's gift to Mary, we must examine all instances of givings until we find one with the donor slot filled by John and the recipient slot filled by Mary. The gift can now be retrieved from the gift slot of giving-1 and we will turn our attention to bicycle-1 and retrieve the required answer from the colour slot of that frame. Compared with semantic nets, the amount of pointer following is reduced, since some of the inference will rather be a matter of matching a given instance frame to a required pattern, getting the required information from within the frame rather than following pointers on to other nodes. Originally this use of frames as templates which could be matched by other frames was given great emphasis, since it was held to provide a means of providing a way of classifying instances according to natural kinds, which it was held were not susceptible to definition by means of necessary and sufficient conditions, as required by a rule-based representation. Latterly, however, this aspect of the use of frames has tended to be somewhat played down, frames being seen rather as a convenient way of grouping and making available static items of information.

6.2.2. Use of frames

The representation we have so far outlined allows us to do all the things we could do with semantic nets. But there is more. Suppose in the bicycle example we had wanted to say that someone likes another person if they give them a gift. This could be expressed as a production rule of the form:

IF (?z donor ?y) AND (?z recipient ?x) THEN (?x likes ?y)

In a semantic net we would have difficulty in representing this. We could draw the appropriate relationship in the general term, but this would not carry through into a specific example of the relation. Similarly in a frame system we would have schemata and slots such as

giving
    donor:
    recipient:
    gift:

person
    owns:
    likes:

etc.

When we instantiate the information we will get

giving-1
    is-a: giving
    donor: John
    recipient: Mary
    gift: bicycle-1

John
    is-a: person
    likes:

Mary
    is-a: person
    likes:

and the information is still not conveyed. What we can do, however, is to attach a piece of code to one of the slots that will force the appropriate conclusion to be drawn. We could thus attach a procedure to the recipient slot of the schema giving, which will place the value of the donor slot of an instance of giving in the likes slot of the frame indicated by the value of the recipient slot of the giving instance. Alternatively, we could attach a procedure to the likes slot of person, so that when an attempt is made to retrieve the value, instances of giving will be searched and the donor of any instance with the frame under consideration as the filler of the recipient slot added to the likes of that frame. Such pieces of code are often termed *demons*, invoked either when a value is put into a slot (often called a *when-added* procedure and analogous to forward chaining in rules) or when a value is requested from a slot (often called a *when-needed* procedure and analogous to backward chaining in rules). The ability to associate the procedures with slots of general terms and pass them on to instances of these terms greatly increases the power of frame representations, since it allows for a dynamic use of information. A frame-based representation will thus have more structure than a semantic net representation since it allows for the incorporation both of attribute values and inferencing methods within the nodes.

6.2.3. Inheritance hierarchies

Within frame systems some kinds of slot can be seen as being of special importance and thus worthy of differential treatment. Any instantiated individual in such a system will be an instance of some general term recognised by the system. Thus, in the example above, John and Mary are instances of the general term person, and bicycle-1 is an instance of the general term bicycle, and this relation is critical, since it will determine the slots present in the instantiated individual and, in the case of some slots, the values also. But this can be taken further. For the general terms themselves can be seen as being examples of more abstract terms. Thus a table is an item of furniture, and an item of furniture is an inanimate object. A dog is a mammal, and a mammal is an animal and an animal is a living thing. In turn, if we wish to pursue the abstraction process, both living things and inanimate objects are physical objects. Just as the general term determines the slots in an instance of that term, so these more abstract general terms determine some of the slots in the more specific general terms. Thus all physical objects have a spatio-temporal location, and so this will apply to both living things and inanimate objects, and hence to animals and items of furniture, and hence to mammals, dogs and tables. This passing of slots down from the more abstract to the more specific is possible because the is-a relation is transitive; thus if a dog is a mammal and a mammal is an animal, then a dog will also be an animal, and so anything that is true of the more abstract term will be true of the more specific term as well.

This gives rise to two possibilities. First, it provides a way of organising our knowledge in a way in which many people find to chime in with their preferences, namely describing a given general term by saying firstly what more abstract terms it exemplifies, and then adding extra information that differentiates it from other general terms falling under that abstraction. Secondly, it means that, given a suitable mechanism, we can make our representation very economical, since we need only represent information at the highest level of abstraction appropriate, and allow this information to apply to more specific terms by being inherited. Considered only in terms of the is-a relation, terms form a hierarchy, which can be conveniently shown in the sort of diagram in Fig. 6.5. Such a diagram is often called a *class hierarchy*, the nodes at a level one higher being called the *superclasses* of those one down, and the immediate descendants of a class being the *subclasses* of that class. For this reason, the is-a relation, when applied to general terms, is often called the *class–subclass* relation, which is helpful in that it avoids confusion with the class–instance relation, which may otherwise be seen as an is-a relation as well.

So useful is this way of looking at the connection between frames that today almost all frame systems will treat this relation differentially and organise the frames in such a class hierarchy, and provide an inheritance mechanism so that slots can be obtained from superclasses when they are not found locally.

In principle this inheritance hierarchy mechanism can be used to apply to other relations as well, so long as they are transitive. Another relation which is often picked out for this kind of treatment is the part-of relation, and a reasonable case can be made out for this, as in the case shown in Fig. 6.6.

It is again a good saving to have only to represent the various components of a head in one place and to allow them to be transferred to anything with a head, for example. This relation is a less obvious candidate, however, for this special treatment, since in this case only some, rather than all, properties will be inherited through the hierarchy. Thus a car will have many attributes that its various parts do not have; cars have wheels, but engines, although part of a car do not. It is thus less clear what inferences we want the inheritance mechanism applicable to the part-of relation to support than was the case for is-a. For this reason, whilst inheritance hierarchies for relations like part-of have been found useful in certain particular applications, they have not found the almost universal popularity of is-a class hierarchies. In the rest of this discussion I shall therefore use inheritance hierarchy to refer exclusively to the latter.

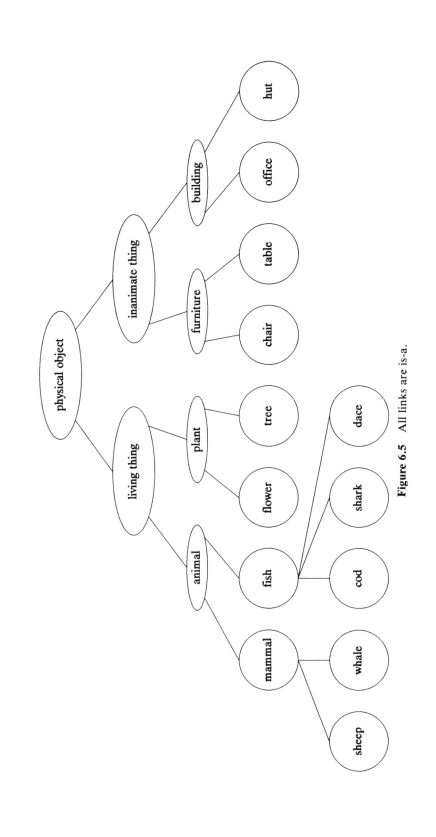

Figure 6.5 All links are is-a.

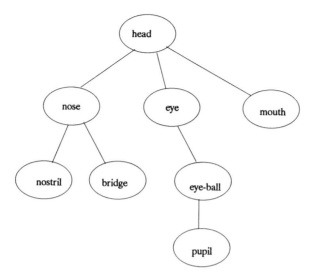

Figure 6.6 All links are part-of.

6.2.4. Default values

In the above discussion we saw that inheritance hierarchies were a good way of representing what slots a given class should have in terms of the classes it was a subclass of, and what was peculiar to itself. We also saw that in some cases it was possible for a class to determine the value of such slots. Thus all subclasses of bicycle will have 2 as the value of its number-of-wheels slot, irrespective of whether it is a racing bicycle, a BMX, a touring bicycle or a mountain bike. Again there is an obvious gain in allowing for this information to be inherited, both in terms of economy of the representation, and in terms of justifying the information: a particular racing bicycle will have two wheels in virtue of its being a bicyle, not in virtue of its being a racing bicycle. In addition, of course, inheriting this information from the superclasses, means that we do not need to be told this information in the case of a particular example. Whenever we hear of a bicycle mentioned, we can take it for granted that it has two wheels, irrespective both of the particular sort of bicycle, and of whether wheels are mentioned in connection with this bicycle. So useful is this feature, which seems to allow us to know considerably more about a situation than has been specifically stated about it, that it is tempting to look for a means of representing as much knowledge as possible in this way. This in turn leads to a pressure to go beyond what we know to be true of all members of a class, and use the same mechanism to record things which are typically, or generally, or mostly true of members of that class.

To take a standard example, it is known that most birds can fly. Thus if someone refers to a bird, then the audience is likely to assume that it can fly unless they are told something to the contrary. It is therefore tempting to use the inheritance mechanism to pass down values for slots which are not confined to values that must be true of all instances of that class, but also values true of most instances of that class. In the above example, the value true will be associated with a flies slot of the class bird, and be unproblematically inherited down to such subclasses as eagles, canaries and robins. Two problems now arise: one is that some of the subclasses will be atypical; thus ostriches, kiwis and penguins do not fly, and the other is that some instances of the class may not fly because of some individual peculiarity; Eddie the eagle may be unable to fly because of a broken wing, or because his feet have been set in concrete (Minsky's example), or because of some other misfortune. If we wish to inherit information of this sort, therefore, we need to complicate the inheritance mechanism by allowing the inheritance of slot values to be blocked, both at the class level, to cope with kiwis, and at the instance level, to cope with the unfortunate Eddie. The problems that can arise here are well discussed in Brachman [10]. A comparison should be made with the production-rule paradigm, and its treatment of this kind of information by giving a general rule and rules governing exceptions to this rule, and reliance on the conflict resolution strategy to sort out the correct answer.

Despite these problems, the use of default values has become extremely popular as a way of representing common-sense knowledge deriving from experience of what is typically the case. Its use requires that reasoning be non-monotonic, in that we will need the ability to withdraw conclusions made on insufficient evidence. This topic will be discussed in detail later in Section 10.3.

6.2.5. Multiple inheritance

One point of debate is whether a class can have more than one superclass. Thus we might see a male person as a class which combined the attributes of maleness, as defined in a superclass, with the attributes of person, also a defined superclass. This can work well, but there are a number of hidden dangers. Suppose that the two superclasses have attributes with different default values. There we must have a means of overriding the unwanted default value. Some systems simply use the order of the appearance of the superclass in the relevant slot, but this can be dangerous: sometimes we will want the default values from one superclass for one slot and the default values from the other superclass for another slot. Sometimes too, slots relevant to one, or both, of the superclasses will not be relevant to the subclass. Thus a toy train is sensibly thought of as being both a toy and a train. But many of the slots we would put in train would be inappropriately inherited by a toy. In radically deviant cases, the whole approach may break down; counterfeit money is not money at all, and a dreamboat is neither a dream nor a boat. Most representation formalisms which permit multiple inheritance avoid these problems by stating what the rules are and leaving the system builder to work around the problems. This again moves away from the possibility of any declarative reasoning, and puts implicit meaning into the representation, which is undesirable.

6.2.6. Object-orientated knowledge representation

A development in the field of programming languages which has had a significant impact on knowledge representation is the increasing popularity of *object-orientated programming languages*. These languages were inspired by the development of Smalltalk [11] in the 1970s, and are increasingly becoming part of the accepted mainstream: witness, for example, the emergence of C + + , which extends C so as to provide object-orientated facilities.

The idea behind these languages is that the program code will consist of a number of definitions of objects, and objects will have associated with them *attributes*, which store values local to the objects, and *methods* which allow operations to be performed on them. These objects are capable of sending messages to one another requesting the performance of one or more of these methods, and a given program will be written in terms of these message-sending operations. Suppose, for example, that we have an object representing a rectangle. This rectangle object will have attributes representing length and breadth, and a method get-area which will return the result of multiplying the one by the other. This multiplication will in turn be effected by sending a message to the object representing the length to multiply itself by the object representing the breadth.

In such a system objects will typically be instances of some class, which provides the definitions of its instances, and the classes will be organised in a class hierarchy to facilitate the inheritance of attributes and methods to lower level classes. This can provide for very effective programming. It is shown to good advantage in the case of writing code for a window management system, as suggested by the following simplified example.

Any window will need to have associated with it certain attributes, such as the location of its bottom left corner, its height and its width, and certain methods, so that it can, among other things, be moved, cleared, re-sized and redrawn. A window with a title will have all these attributes, together with some peculiar to itself, such as the text of the title. It will also need all these methods, some of them being specialised to allow for operations associated

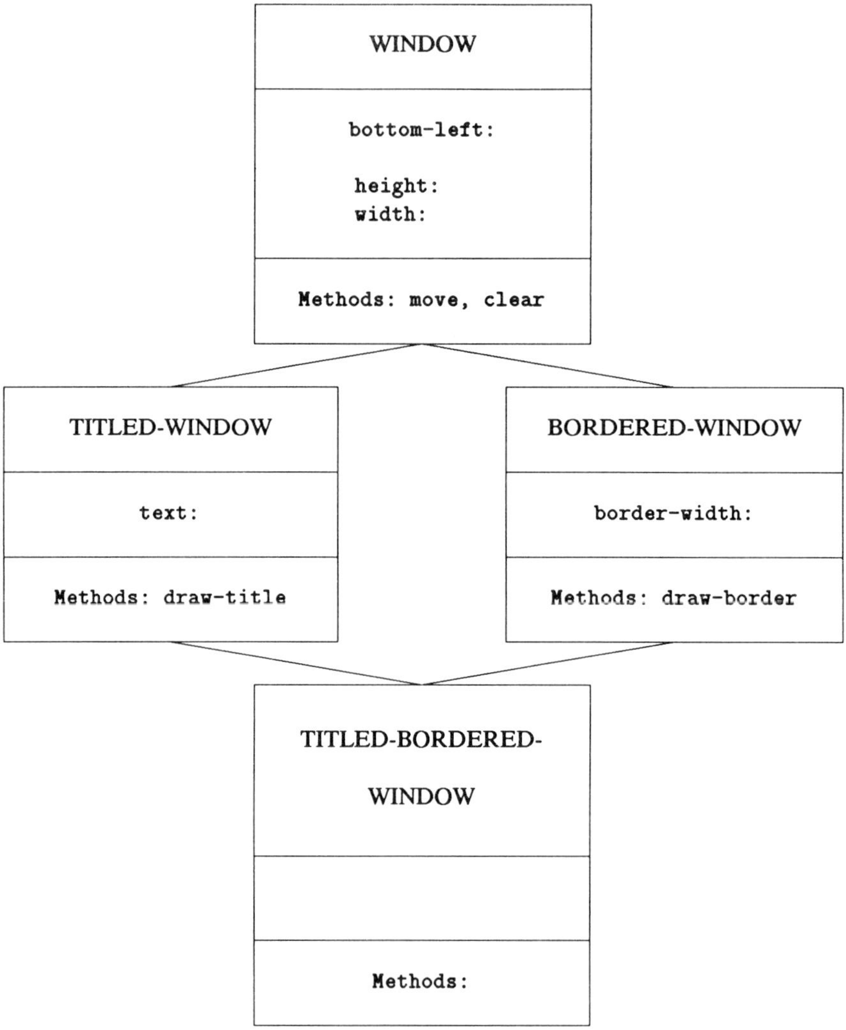

```
┌─────────────────────────────┐
│          WINDOW             │
├─────────────────────────────┤
│      bottom-left:           │
│      height:                │
│      width:                 │
├─────────────────────────────┤
│   Methods: move, clear      │
└─────────────────────────────┘
```

```
┌─────────────────────┐         ┌─────────────────────┐
│   TITLED-WINDOW     │         │  BORDERED-WINDOW    │
├─────────────────────┤         ├─────────────────────┤
│      text:          │         │   border-width:     │
├─────────────────────┤         ├─────────────────────┤
│ Methods: draw-title │         │ Methods: draw-border│
└─────────────────────┘         └─────────────────────┘
```

```
┌─────────────────────────────┐
│     TITLED-BORDERED-        │
│        WINDOW               │
├─────────────────────────────┤
│                             │
├─────────────────────────────┤
│        Methods:             │
└─────────────────────────────┘
```

Figure 6.7

with the title. Similarly a window with a border will need the general window attributes, together with some peculiar to itself, such as border width, and the general methods, some customised so as to allow the display of the border. A window with both a title and a border will need all the attributes and methods proper to both windows with titles and windows with borders. All of this can be conveniently implemented by a hierarchy of the following sort in Fig. 6.7.

This style of programming incorporates a number of widely advocated features. It has the advantage of economy, in that common elements are defined only once. It supports data abstraction in that a message can be sent to a window object without any need to know which sort of window it is, in the knowledge that the method appropriate to the right sort of window will be executed. It supports modularity in that an object includes within itself all its local variables and processing methods. Finally it exhibits the property of data hiding, and all the destructive assignments and access to local variables are internal to an object and require sending the appropriate message to the object.

The essence of problem solving within the object-orientated paradigm is to identify the real-world objects relevant to the problem, the attributes of those objects, and the processing operations in which they participate. Writing a program then consists of defining those objects, and instantiating them in a simulation. Sending the appropriate message will then cause the simulation to execute so as to solve the problem. Abstract data types, modularity, inheritance and dynamic type binding, all prominent features of object-orientated languages, make this style of problem solving convenient and effective.

The development of object-orientated programming languages opened up exciting possibilities for knowledge representation. The similarities with frame-based class hierarchies are evident, since the objects in the object-orientated system look very like the frames in a frame system, and the various procedures attached to the slots in the frame system are not dissimilar to the methods of the object-orientated system. It therefore seems an attractive notion to import wholesale the ideas from object-orientated programming into knowledge representation, and to use an object-orientated system such as Smalltalk as a knowledge representation language, and so get a knowledge representation language for free.

This is a highly attractive idea, combining as it does both developments from the theory of programming languages, and knowledge representation. In practice, however, it has been found that the use of methods can become highly procedural, and difficult to reconcile with the desire for knowledge representation to be as declarative as it can be. Although destructive assignments are hidden within objects, they none the less occur. Systems using object-orientated methods have proved highly effective for certain applications where simulation fits naturally, such as the design of electrical circuits and the control of chemical processes. In these sorts of applications the destructive assignment inherent in keeping over-writable values in the attributes of objects is found to overcome many of the problems associated with time (the fact that the relationships a given object stands in may vary as time goes by; for instance the temperature of a given mixture in a chemical

98 Knowledge Representation

process may well be in a constant state of flux). In such cases the procedural nature of methods and the destructive assignment involved in over-writing slot values have positive advantages. For such areas therefore, the representation has much to commend it.

It does not, however, overcome the criticisms that were levelled at structured object representations earlier, such as those made by Brachman and Woods, concerning the use of default values, and the precise import of a class definition. Moreover in some domains it is very hard to identify the appropriate objects and the best way of organising them in a class structure. Whilst some domains naturally lend themselves to a simulation approach, others do not. The conclusion must be that object-orientated languages do provide a good, natural means of implementing structured object representations, and this makes such representations more attractive to use in practice, but they do not do anything to obviate the problems that such representations encounter. Thus object-orientated languages add little to the theory of knowledge representation, and their use for knowledge representation is simply an exemplification of the structured object paradigm. The undoubted possibilities of object orientation, considered as a programming paradigm, should not lead us to see more to it than there is from a knowledge representation standpoint.

6.2.7. Frames combined with rules

Relatively recently, as AI and its techniques have started to form a standard repertoire, there has been a great interest in combining the various forms of knowledge representation. In particular, it has been sensed that the attractions of representing some kinds of knowledge in a class hierarchy are entirely independent from the attractions of representing other information in production rules. It will be recalled from Section 5.2.1 that the working memory of production-rule systems was often realised in the form of entity–attribute–value triples, and that semantic nets were also based on a notational variant of this formalism, and that the frame-based formalism required special-purpose procedures to relate the attributes of frames within the system. From this it is only a small step to the idea that the facts relating to a situation, which we might call the static information, would be well represented in a frame/semantic net formalism, whilst the dynamic information which related the values of different slots might be better represented as production rules. What is proposed here is that the procedures attached to slots should not be procedural code, but rather some declarative expression of the inter-slot relationships.

One of the earliest systems to attempt to combine frames and rules, and which provides a good example of such a system is the CENTAUR system

[12]. This system is a re-implementation of PUFF, an expert system to diagnose lung disorders originally implemented in EMYCIN, so as to remedy a number of knowledge representation defects felt in the original PUFF. These defects included that of representing typical patterns of data, of modifying the rule base, and generating clear explanations. By using frames as well as production rules it was hoped to provide a structure which would mitigate these problems. This system was largely successful, but the use made of frames was mainly structural, for packaging knowledge into more easily applicable units, rather than anything representationally fundamental. This is perhaps explained by the fact that the starting point was a production-rule system, not a frame-based system.

6.2.8. Combining frames with logic—KRYPTON

Another interesting hybrid representation scheme is KRYPTON, developed by Brachman, Fikes and Levesque [13]. It is, in part, an attempt to produce a representation which has the benefits of structured object representations for describing complex entities, whilst not falling foul of the kinds of criticisms made by Brachman of the standard use of the representations.

KRYPTON's representation contains two components, known as the *TBox* and the *ABox*. The TBox provides a means of constructing complex structured descriptions, which will form the terms used to make assertions about what is the case using the ABox. The TBox language permits expressions for concepts, which correspond closely to frames, and for roles, which correspond to slots in frames. These slots, cannot, however, be accessed directly, which means that these concepts cannot serve as a repository for assertions. The problem of necessary and sufficient conditions for natural kinds is handled by a type of specialisation which allows all members of the subclass to be members of the superclass but which have no sufficient conditions for determining membership of the subclass. Thus membership of the subclass cannot be deduced, only specifically asserted. This is a sound principle, but somewhat conservative. The ABox is standard first-order predicate calculus, except that the basic non-logical symbols are terms defined in the TBox. Thus the thrust of the scheme is to use the TBox to define a language which can then be used to make assertions with the ABox.

KRYPTON is not currently in a state which would permit the development of substantial applications. None the less it is a highly interesting tool for exploring a number of research ideas, especially, from our perspective, the recognition of the different representational needs of definition and assertion and the reflection of those different needs through a difference in formalism. Whether the approach taken is one which is capable of application to all domains, since the role of definition is less important in some than in others, is still an open question.

6.2.9. AI toolkits

The obvious attractions of combining styles of representation have led to the development of what are called AI toolkits, well-known examples of which are Knowledgecraft, produced by the Carnegie group, and KEE, produced by Intellicorp. These toolkits offer an environment, designed to run on a high-performance workstation, for constructing knowledge-based systems: usually the means of building a class hierarchy, and the facility to attach methods, written in LISP, to these classes, and some sort of rule system, whether based on a production-rule language or on a logic programming language such as PROLOG. In addition, a variety of sophisticated editing, debugging and interface tools are provided. These toolkits have become very popular and are used in many current research projects both in academia and industry.

References

1. Nilsson, N.J., *Principles of Artificial Intelligence*, Springer-Verlag, Heidelberg, 1982.
2. Jackson, P., *Introduction to Expert Systems*, Addison-Wesley, Reading, Mass., 1986.
3. Quillian, M.R., "Semantic memory", in *Semantic Information Processing* (ed. M. Minsky), MIT Press, Cambridge, Mass., 1968, pp. 216-70. The work also formed Quillian's Ph.D. thesis, Carnegie Institute of Technology, 1966.
4. Winston, P.H., "Learning structural descriptions from examples", in *Psychology of Computer Vision* (ed. P.H. Winston), McGraw-Hill, New York, 1975, pp. 157-209.
5. Schank, R.C., "Conceptual dependancy: A theory of natural language understanding", *Cognitive Psychology*, 3, 552-631 (1972).
6. Woods, W.A., "What's in a link?: Foundations for semantic networks", in *Representation and Understanding: Studies in Cognitive Science* (ed. D. Bobrow and A. Collins), Academic Press, New York, 1975, pp. 35-82.
7. Brachman has written several papers on this topic. The most accessible is Brachman, R.J., "I lied about the trees", *AI Magazine*, 6(3), 60-93 (1986).
8. Quillian, *op. cit.*
9. Minsky, M., "A framework for representing knowledge", in *The Psychology of Computer Vision* (ed. P.H. Winston), McGraw-Hill, New York, 1975, pp. 211-277.
10. Brachman, R.J., "On the epistemological status of semantic networks", in *Associative Networks: Representation and Use of Knowledge by Computers* (ed. N.V. Findler), Academic Press, New York, 1979, pp. 3-50.
11. Smalltalk was a language developed in the early 1970s at Xerox Palo Alto Research Center. It is now widely available. A good discussion of the language

and its concepts, and of object-orientated programming in general can be found in Sebesta, R.W., *Concepts of Programming Languages*, Benjamin/Cummings Publishing Company, Redwood City, California, 1989.
12. Aikins, J., "Prototypical knowledge for expert systems", *Artificial Intelligence*, **20**, 1983, pp. 163–210.
13. Brachman, R.J., Fikes, R.E. and Levesque, H.J., "KRYPTON: integrating terminology and assertion", *Proceedings AAAI-83*, Morgan Kaufmann, Los Altos, California, 1983.

7
Logic and Predicate Calculus

In this chapter we shall examine the third of the major knowledge representation paradigms—logic—and in particular the first-order predicate calculus. A number of logics have been developed in philosophy and mathematics to represent arguments and to assess their soundness or unsoundness. This in turn has meant that they have developed as a means of representing the premises of arguments and their conclusions, and this is sufficiently close to the idea of a knowledge base and the inferences which can be drawn from it to give hopes of transferring the representational ideas developed in logic in a relatively straightforward manner. The simplest form of logic is the propositional calculus, but this represents only the logical relations between whole statements, and its expressive power is therefore somewhat limited. Despite these limitations, several popular expert system shells can be seen as being based on the propositional calculus, showing that this simple form of logic has a place in AI. More expressive is the first order-predicate calculus, which expresses relations between objects, asserting and denying these relationships, and stating the logical relations between these statements. Some have found the expressiveness of this form too limited for their purposes, however, and have developed logics to handle temporal or deontic (relating to permission and obligation) relations as well. Finally one may distinguish classical logics, in which a statement always has one and only one of two possible truth values, truth or false, and has its truth value unchangeably, from what are sometimes called "exotic" logics, such as polyvalent logics, where statements may have one of several truth values, intuitionistic logics, which permit statements to lack truth value, and non-monotonic logics in which the truth value of a statement may change. Despite this plethora of logics, however, and despite the fact that any could be used as the basis of a knowledge representation, the first-order predicate calculus is the one which is usually intended when logic is proposed as a knowledge representation paradigm.

7.1. Advantages of predicate calculus

The reason why many people are attracted to predicate calculus as a knowledge representation formalism is well expressed by Hayes [1]:

> I propose . . . that first-order logic is a suitable basic vehicle for representation. However, let me qualify this. I have no particular brief for the usual syntax of first-order logic. Personally I find it agreeable: but if someone wants to write it all out in KRL [a structured object representation] or semantic networks of one sort or another . . . well, that's fine. The important thing is that one knows what it means: that the formalism has a clear interpretation. . . . At the level of representation, there is little to choose between any of these, and most are strictly weaker than predicate calculus, which also has the advantage of a clear, explicit model theory, and a well-understood proof theory.

Here Hayes is extolling predicate calculus as a formalism on the grounds that it is at least as metaphysically and expressively adequate as any other formalism but that it has advantages over others in its lack of ambiguity and its clarity. Notational convenience is a matter of personal taste, and computational aspects are outside his concern, although it would, if you accept his choice of notation, be highly desirable if it were also computationally adequate. He goes on:

> in claiming equivalence [between predicate calculus and other paradigms], one is speaking of representational (expressive) power, not computational efficiency.

Of course, computational tractability is an issue when putting the representation to work in an AI system, and we shall see how this has forced those keen to use predicate calculus when building AI systems to restrict themselves to a subset of full first-order logic in order to obtain the required degree of computational tractability. We shall need also to consider what expressive power is sacrificed by this restriction. For the moment, however, we need to recognise that predicate calculus is a natural choice for knowledge representation because it was developed specifically in order to give a precise and unambiguous representation of statements, and because it offers a clear view of how these statements should be manipulated in its deductive proof theory.

Hayes is primarily interested in representing knowledge, rather than in building a program to manipulate it. The popularity of first-order logic in AI has, however, been greatly enhanced by those who have seen in logic and logical deduction a paradigm for the entire activity of programming. This insight led to the development of *logic programming*, in which a program is seen as a set of axioms expressed in first-order logic and the execution of the program as the deduction of consequences from those axioms. This gives rise to the highly attractive notion of writing computer programs simply by representing the knowledge pertinent to the problem to be solved. In the remainder of this chapter we will concentrate on logic programming and focus on logic as an executable representation, rather than as a representation *per se*.

7.2. Foundations of logic programming

The whole enterprise of logic programming is founded on the ability to produce an interpreter which will deduce consequences from a set of logical axioms. In the early days of AI there was much interest in building automated theorem provers which attempted to do just this. Although much good and interesting work was carried out, none of these theorem provers performed with sufficient speed to be considered as the basis of a programming language interpreter. Logic programming therefore had to wait for a breakthrough in this field, which came with the discovery of the *resolution* principle by Alan Robinson [2]. The key here is that systems of natural deduction reflect the preference of people for a multiplicity of deduction rules. Rather than transform expressions into a normal form, they prefer to work with shorter, more varied, and more natural, expressions and select the inference rules to apply to form the proofs they need from a relatively large set. This requires the exercise of considerable judgement, and this aspect of the theorem proving process was hard to automate. Mechanical theorem proving needed, if it was to be general-purpose, in contrast to natural deduction, as much uniformity, both in syntax and in deductive rules, as possible. What was needed therefore was a single rule which could be applied to normalised expressions without judgement, but which would subsume the various natural deduction rules, so that completeness would not be lost. Resolution is such a rule, and is fundamental to the use of predicate logic as a general-purpose knowledge representation paradigm. It should be stressed, however, that resolution methods are concerned with the production of a general-purpose theorem prover. When the goal is to prove substantial theorems in a particular domain, such as set theory, the use of non-resolution methods together with domain-specific heuristics has met with some success. A good discussion of non-resolution theorem provers can be found in Ref. 3.

7.2.1. Resolution

There are three varieties of normal form: *conjunctive normal form*, which comprises a conjunction of disjunctions (product of sums), *disjunctive normal form*, comprising a disjunction of conjunctions (sum of products), and *clausal form*, in which a disjunction is said to be implied by a conjunction. Any well-formed expression of first-order logic can be translated into any of these normal forms, and there is, as was shown in Section 3.3.8, a straightforward translation between them. Resolution can be applied to any of the normal forms. In the discussion here we will mainly use clausal form, as that is the normal form of the logic programming language PROLOG, and because it has the closest correspondence to the production-rule formalism discussed in Chapter 5.

First, however, we will see how resolution applies to conjunctive normal form (CNF), since the rationale for resolution is somewhat clearer in this case. Suppose we have the following expression in CNF:

CNF1 (P ∨ Q ∨ – R) & (S ∨ – Q ∨ T)

It will be noticed that one of the literals, Q, occurs in both the disjunctions, positively in the first disjunction and negatively in the second. What resolution permits us to do is to resolve these two disjunctions on their common literal so as to produce a single new disjunction, called the *resolvent*:

CNF2 P ∨ – R ∨ S ∨ T

The resolvent is a logical consequence of the original expression. The idea underlying resolution is that by successively resolving pairs of clauses we can derive all the consequences of the original set of axioms. Resolution is a powerful proof rule, but is somewhat unintuitive. It is therefore worth giving an informal explanation of how it works. Consider CNF1 again. It contains a literal Q which we know must either be true or false (since in classical logic it has one and only one of the truth values true or false). Suppose that it is true: then from the second disjunction we can conclude S ∨ T (since one of S, – Q and T must be true). Suppose that it is false: now, in a similar fashion, we know that one of P and – R must be true. But Q must be either true or false, so, combining the two cases, we get the resolvent P ∨ – R ∨ S ∨ T.

Now consider resolution as applied to clausal form. To convert from CNF to clausal form we disjoin the positive literals on the left-hand side of the implication and conjoin the negative literals on the right-hand side. Thus CNF1 becomes the pair of clauses:

CL1 P ∨ Q ← R

and

CL2 S ∨ T ← Q

In clausal form the literal to resolve on must appear on the LHS of one clause (a positive occurrence in CNF) and on the RHS of another clause (a negative occurrence in CNF). The resolvent is simply produced by combining the LHSs and RHSs of the two clauses and omitting the common literal: thus in the example the resolvent is

$$\text{CL3 } P \vee S \vee T \leftarrow R$$

Note that only one literal can be resolved on at a given time:

$$\text{CL4 } P \vee Q \vee R \leftarrow S$$

and

$$\text{CL5 } U \vee W \leftarrow Q \,\&\, R$$

do not resolve to give

$$\text{CL6 } P \vee U \vee W \leftarrow S$$

since CL4 and CL5 are true and CL6 false if S and Q are true and P, U, W and R are all false. The possible resolvent

$$\text{CL7 } P \vee Q \vee U \vee W \leftarrow Q \,\&\, S$$

is true, but vacuous, since Q trivially implies Q.

It is worth looking now to see the sense in which resolution subsumes some of the more familiar rules of natural deduction. Consider first modus ponens:

$$\frac{\begin{array}{l} P \rightarrow Q \\ Q \end{array}}{Q}$$

Expressed in clausal form this becomes

$$\begin{array}{l} Q \leftarrow P \\ P \leftarrow \end{array}$$

We can resolve on P to get $Q \leftarrow$. Next consider modus tolens:

$$\frac{\begin{array}{l} P \rightarrow Q \\ -Q \end{array}}{-P}$$

In clausal form we have

$$\begin{array}{l} Q \leftarrow P \\ \Box \leftarrow Q \end{array}$$

where $\Box$ is used to represent absurdity so that $\Box \leftarrow Q$ can be taken as a denial of Q. We can resolve these two clauses to get $\Box \leftarrow P$, which is the denial of P. As a more complicated example we can consider disjunction elimination:

$$P \vee Q$$
$$P \rightarrow R$$
$$Q \rightarrow R$$

$$R$$

In clausal form this becomes

$$P \vee Q \leftarrow$$
$$R \leftarrow P$$
$$R \leftarrow Q$$

We can resolve the first two clauses to get R ∨ Q ←, and resolve this resolvent with the third clause to get R ←. The remaining natural deduction inference rules are left for the interested reader to consider.

7.2.2. Control of general resolution

In the very small examples considered above there were in each case only one or two resolutions that could be performed. This meant that there was little difficulty in selecting pairs of clauses to resolve, and that the number of consequences that could be generated from the clauses was small. In a realistic example, however, there may well be a large number of potential resolvents, and the choice made will crucially effect the efficiency with which a desired consequence is deduced, and the general behaviour of the system. We therefore need some strategy which will assist us in proving the consequences we require.

In the examples above we started from the initial set of clauses and deduced consequences until we found the consequence we were interested in. This is, of course, analogous to the use of data-driven reasoning using production rules, and suffers from some of the same problems, in that there is no mechanism to ensure that the chain of inference is leading towards the desired consequences, and so many unneeded consequences may be produced. In a resolution context, this strategy is often called *bottom-up* resolution. There is, however, a strategy similar to goal-directed reasoning, often called *top-down* or *refutation* resolution, and applicable when there is a particular consequence that is required, and which can serve as the goal. To use this strategy, the negation of the desired consequence is added to the initial set of clauses and resolution proceeds until the empty clause (□ ←) is obtained. If the empty clause is obtained, this means that a contradiction can be derived from the initial set together with the negation of the goal. This in turn means that if the initial clauses are all true, then the negation of the goal cannot be true. In classical logic this means that the goal must be true, and so the goal is provably a consequence of the initial set. Consider the following example:

> CL8 P ← Q
> CL9 Q ← R & S
> CL10 R ←
> CL11 S ←

Suppose we wish to prove P is a consequence of CL8–11. Add the negation of P:

> CL12 □ ← P

Resolve CL8 and CL12 to get

> CL13 □ ← Q

Resolve CL9 and CL13 to get

> CL14 □ ← R & S

Resolve CL10 and CL14 to get

> CL15 □ ← S

Finally, resolve CL11 and CL15 to get the empty clause

> CL16 □ ←

We now have a contradiction form □ ← P, and hence a proof of our goal, P ←.

The successive chain of denials generated by this strategy makes it possible to focus the search on the clause we are interested in, but we still need to make sensible choices as to which clauses to resolve within the overall refutation resolution strategy. At any given time there will be a number of potentially resolvable clauses, and we need guidance in the selection of the most appropriate pair. Thus just as a production-rule interpreter required a conflict resolution strategy to select a rule for firing, so too an automated theorem prover employing resolution will require a resolution strategy to determine which clauses should be resolved.

A number of such strategies have been proposed, and I shall mention some of them here. First, we may consider the equivalent of breadth-first search. This strategy demands that all resolutions possible from the initial set of clauses are performed before any of the resulting resolvents are themselves resolved. As one might expect, given the properties of breadth-first search search, such a strategy is complete, but grossly inefficient. Second, there is what is known as *set of support*. In this strategy each resolution, after the first, will use at most one clause from the initial set of sentences. This strategy is also complete, and in practice will find a given solution more quickly than a strategy using breadth-first search. This can be easily seen in the case of top-down resolution, since choosing a derived clause as one of the clauses to resolve will ensure that we always have □ as the LHS, and so have potentially the empty clause as resolvent. A third possible consideration is called *unit preference*. This strategy will use clauses with a single literal wherever possible. Use of such clauses will obviously produce resolvents with fewer clauses, and so tend to focus search towards the generation of empty clauses, a clear advantage when using a refutation strategy. Finally we may mention *linear input*, which insists that at least one of the clauses being resolved comes from the initial set of sentences. This strategy is of particular interest since, whilst it is not complete for full clausal form, it is complete, simple and efficient if the initial clauses are all Horn Clauses. It is this strategy which, combined with set of support, forms the basis of PROLOG. For a full discussion of resolution strategies, see Ref. 4.

One important difference between the resolution strategies mentioned here and conflict resolution in production-rule systems should be emphasised. The production system generates a conflict set of all rules that can be fired, whereas in resolution a single clause is selected. This is useful since we can avoid worries about the efficiency of maintaining the conflict set from cycle to cycle.

7.2.3. Horn Clauses

This last section mentioned Horn Clauses, which are a central notion in logic programming. Horn Clauses are a subset of first-order predicate calculus, in that they are sentences which are in clausal form but which contain at most one literal on the LHS. Thus the following are all Horn Clauses:

HC1 $P \leftarrow Q \ \& \ R$
HC2 $P \leftarrow$
HC3 $\Box \leftarrow P$
HC4 $\Box \leftarrow$

Whereas the following are not:

NHC1 $P \lor Q \leftarrow R$
NHC2 $P \lor Q \leftarrow$

The key restriction as compared with full clausal from is that it is not possible to express disjunctive conclusions. It should also be noted that negated literals do not exist in Horn Clauses, since transformation into clausal form eliminates negations by putting the literals on the appropriate side of the implication, as was shown in Section 3.3.8. Horn Clauses are therefore a subset of full clausal form, and as a consequence are less expressive. They are important in logic programming, however, because they are so much more computationally tractable than full clausal form that it is possible to build a theorem prover for this subset which is sufficiently efficient to serve as a basis for a practical logic programming language. There is thus a strong incentive to sacrifice the expressiveness of full clausal form to achieve these computational gains. Whether the sacrifice is too great will be considered in detail in Section 10.2.

7.2.4. Resolution and predicate calculus

Although I spoke in terms of predicate calculus when introducing resolution in the previous section, all the examples were confined, for simplicity of exposition, to the propositional calculus. The problems that are created when we introduce variables arise from the need to determine whether or not two literals in different clauses containing variables match or not, so as to decide whether we can resolve those clauses. Additionally there is a complication in putting sentences containing variables into normal form. The latter problem is solved by a method known as *Skolemisation*, and the former by a process called *unification*. We need to look at both these topics in turn.

7.2.5. Skolemisation

When putting an expression of the predicate calculus into CNF, we begin by eliminating biconditionals and conditionals and driving in negation just as for propositional calculus, as described in Section 3.3.8. We will then be left with an expression such as

E1 (x)(Fx) & (∃x)Gx

The first thing we must do is to give all the variables in the expression occurring within the scope of different quantifiers distinct names. Thus

E1.1 (x)Fx & (∃y)Gy

We next eliminate the existential quantifiers by Skolemisation, which means first that we replace existentially quantified variables not within the scope of universal quantifiers by Skolem constants, which are simply arbitrary names. So in the example we would get

E1.2 (x)Fx & Ga

where ''a'' is simply an arbitrary name not occurring elsewhere in the expression. We are able to do this because there must be something which is G, and there is no reason why we should not choose to refer to that thing, whatever it is, as "a". There is, however, a complication if the existentially quantified variable occurs with the scope of a universal quantifier. Consider the following representation of "everyone has a father":

E2 $(x)(\exists y)(Fyx)$

If we were simply to replace the existentially quantified variable by an arbitrary name, to get (x)Fax, we would in effect be saying that something, which we choose to call a, is the father of everyone, which represents

E3 $(\exists y)(x)Fyx$

and not expression E2 at all. What we have to record is the fact that the value of y depends on the value of x, and we do this by means of a *Skolem function*, an arbitrary function name which takes an individual within the domain as argument and maps that individual on to the value of another individual in the domain. So we write E2 as

E2.1 $(x)Fg(x)x$

where g(x) is a Skolem function of x.

This means that it is possible for y to take different values depending on the value of x, which is what we required for E2. One might be a little worried that the meaning of the expressions concerned is being traduced by the substitution of these arbitrary names, but we can offer the following justification of Skolemisation, namely that there is an interpretation of the symbols of the original expression which makes that expression true if and only if the Skolemised version is true. The expressions are, therefore, in this sense, equivalent.

Once we have eliminated the existential quantifiers in this way, we can convert the expression to *prenex form* by simply moving the universal quantifiers to the beginning of the expression. Thus

E4 $(x)(Fx \& (y)(Rax \& Ray))$

is equivalent to

E4.1 $(x)(y)(Fx \& Rax \& Ray)$

Now we can simply drop the quantifiers and all free variables in the expression will be implicitly universally quantified. Now disjunction can be distributed over conjunction, as for propositional calculus and we will have the expression in CNF with literals which are either predicates with a constant term or a variable, and negations of these (e.g. Fa, Fx, − Fa, − Fx) or relations containing constants, variable or both and negations of these (e.g. Rab, − Rab, Rxy, − Rxy, Rax, − Rxa, etc.). We easily convert such an expression to clausal form by gathering the literals onto the appropriate sides of the implication, as in the propositional case.

7.2.6. Unification

We can now turn to the problem of deciding whether or not two literals, one or both of which contain one or more variables, match. Consider Fa and Fx. Intuitively we can say these match just in case that the value of x is restricted to a. We know that a is a possible value for x, since Fx is universally quantified, so we can impose this restriction, although it must be imposed consistently with regard to any other references to x. Similarly we can see that Fx will match with Fy just in case that the same value is substituted for both x and y. This could be achieved by substituting a constant, say a, for both x and y, or replacing x by y. In contrast Fx can never match with Gy since the predicates are different, and Fa cannot match with Fb, since the constants are different. These considerations give a reasonable informal account of what we want, but we should see how we can be more formal.

Given an expression containing variables we can define a *substitution* (often written as θ) as a set of assignments of terms to variables such that no variable is assigned more than one term. Applying a substitution to the expression yields a *substitution instance* of the expression, which we write $E\theta$. For example, given the expression,

$$Fxy \leftarrow Gyz$$

and the substitution

$$\{x := a, y := b, z := c\}$$

we get the substitution instance

$$Fab \leftarrow Gbc$$

whereas the substitution

$$\{x := y, z := y, u := a\}$$

would give the substitution instance

$$Fyy \leftarrow Gyy$$

We can now say that a *unifier* for two expressions, E1 and E2, is a substitution which when applied to the expressions yields identical substitution instances. So that, given the expressions

$$Fxy$$

and

> Fwz

we can propose the unifier

> {x: = a, y: = b, w: = a, z: = b}

which will give the common substitution instances

> Fab

An alternative unifier would be

> {w: = x, z: = y}

giving the common substitution instances

> Fxy

7.2.7. The most general unifier

From this we can see that there will often be a large number of possible unifiers for a given pair of literals. Given this choice, it is necessary to choose the unifier that will be most useful to us. We therefore introduce the notion of the *most general unifier*. This is essentially the unifier which gives rise to least commitment and assigns only those variables, and only to the extent necessary to yield a common substitution instance, which is to say to match the two expressions. The most general unifier will therefore be the one which makes the least specific assignments. Thus, given expressions

> Fx and Fy

the unifier

> {x: = y}

will be more general than

> {x: = a, y: = a}

because in the latter case the variables are specifically assigned to a particular value, whereas in the former case the only restriction is that they take the same value, whatever that may be. Formally we may say that θ is the most general unifier of E1 and E2 if and only if

(1) θ is a unifier of E1 and E2, and
(2) the common instance E1ψ determined by any other unifier ψ of E1 and E2 is an instance of the common instance E1θ determined by θ, so that E1ψ = (E1θ)λ for some substitution λ.

Now that we have defined a notion of unification, we can consider the effect of using resolution on clauses containing variables and employing unification to match literals to resolve on. Each time a unification is made we make an assignment restricting the possible values of the variables involved in the clauses resolved upon. Thus what we get when we prove a given goal is a model for that goal, namely a set of substitutions for the variables in the goal expression, which exemplify the truth of the goal expression, given the axioms. In the case of bottom-up resolution this model is achieved directly, whereas in top-down resolution it is found by finding a counter-example to the negation of the goal expression. This seems to give us the sort of theorem prover we want for logic programming, and one which can perform efficiently enough to form the basis of a practical programming language, provided we restrict ourselves to expressions that are part of the Horn Clause subset.

7.2.8. An example

Let us consider another example to see how this works in practice. Suppose we have the following set of axioms consisting of a single rule and a number of facts:

LP1 Fxy ← Gx & Hy
LP2 Ga ←
LP3 Gb ←
LP4 Gc ←
LP5 Hb ←
LP6 Hc ←

We can now use top-down resolution to prove, for example $(\exists x)(\exists y)(Fxy)$. First we can re-write the goal using universal quantifiers to get $-(x)(y)(-Fxy)$, which goes into clausal form as Fxy ←. We therefore add the negation of this goal expression:

LP7 □ ← Fx'y'

Now we can resolve LP7 with LP1 to get

LP8 □ ← Gx & Hy   with unifier {x': = x, y': = y}

which resolves with LP2 to give

LP9 □ ← Hy   {x: = a}

which resolves with LP5 to give

LP10 □ ←   {y: = b}

Substituting the constants found through unification in the original goal, we get □ ← Fab. Thus Fab is a counter-example to the negation of the original goal, and so a model for that goal. There are, of course, other counter-examples implied by the axioms, and a different choice of clauses to resolve would have found different ones. Thus if we had resolved LP8 with LP3 the unifier would have been {x: = b}, and the counter-example found would have been Fbb. Making different choices in this way would have produced all the models (Fab, Fbb, Fac, Fbc, Fcb and Fcc) for the original goal, and, of course, none of the non-models such as Fba.

This example indicates that the choice of clauses to resolve, whilst not affecting whether or not the goal is proved, is important in that different choices may produce different models for the goal. Of course, we would wish to be able to get the complete set of models from the program, but it may be important to order them in some way. Thus the resolution strategy we use must make its choices in a systematic way, so that we can predict what order the models will be found in, and must also be capable of eventually making all possible choices, so that all models will eventually be found. A further desirable property is that a model satisfying the goal should be found as quickly as possible, since we may often be interested primarily in whether the goal is true, rather than in any particular model.

One method of achieving this, of particular interest since it is the method adopted by PROLOG, the paradigmatic logic programming language, is to use depth-first search and to backtrack to the most recent choice point on termination, whether the termination results from success or failure. Within clauses, literals are selected in the order in which they appear in clauses, and clauses themselves are selected in the order in which they appear in the database. This achieves the aims outlined in the previous paragraph, the only problems arising when a path fails to terminate, as this will mean that no more models will be found.

7.2.9. Procedural semantics for Horn Clauses

Another important point about logic programs restricted to Horn Clauses is that they can be given an operational character by reading them procedurally as well as declaratively, as above. Essentially the procedural reading involves seeing goals as sets of procedure calls which are processed by calling the appropriate procedure. In this interpretation, a resolution step is a procedure calling operation which selects a procedure, performs unification, replaces the original call by the body of the selected procedure, and applies the unifier to the result. It was the elaboration of the procedural semantics of logic programs which helped to give a good deal of credibility to logic as a programming language. It is inappropriate to give full details here; the interested reader may find them in Hogger [5]. However, a simple example will suffice to show the general principles. Suppose we have a logic program:

   PI1 Fxy ← Gx & Hxy
   PI2 Gz ← Jz
   PI3 Ja
   PI4 Hab

and we wish to solve the goal Fx'y'.

First we will call Fx'y' as a procedure. This call will be processed by selecting PI1 with unifier {x: = x', y: = y'}. We may now replace the original procedure, Fx'y', by the body of the selected procedure, Gx & Hy, and apply the unifier to leave the set of procedures to be called, [Gx', Hx'y'].

Next we will call Gx', so we can select PI2, using as unifier {z: = x'}. Replacing the original call by the body of the new call and applying the unifier we get [Jx', Hx'y'] as the procedures to be called.

The call to Jx' will lead us to select PI3 with unifier {x': = a}. The body of PI3 is empty so that we get, when we apply the unifier, [Hay'] as the only remaining procedure call.

This call can be effected by selecting PI4 with {y': = b} as unifier. Thus the original procedure call succeeds with x' = a and y' = b.

7.3. Main ideas of logic programming

Kowalski [6] says:

> Conventional algorithms and programs expressed in
> conventional programming languages combine the
> logic of the information to be used in solving
> problems with the control over the manner in
> which the information is put to use. This
> relationship can be expressed symbolically by the equation
>
>   Algorithm = Logic + Control.

This paragraph neatly encapsulates the fundamental idea of logic programming; namely that given this programming style it should be possible to produce a program simply by stating the declarative relationship between inputs and outputs, leaving control to the theorem prover which can execute this problem statement, and so freeing the programmer from detailed considerations as to how this relationship is to be effected. This view means in effect that programming is reduced to knowledge representation.

7.3.1. Logic programs as executable specifications

As an example, consider the problem of appending a list to another list to give a third list. We will have two lists as input, say [a,b,c] and [d,e,f], and the output will be another list, in this case [a,b,c,d,e,f]. The obvious way of achieving this would be successively to take terms off the second list and put them on the end of the first list. This is, however, not the easiest implementation, since it is easier to push a term on to the front of a list rather than the end. Thus we might suggest instead taking terms off the first list and pushing them onto a temporary stack, until there are no more terms in the first list. We can then pop the terms off the stack and push them onto the second list to give the third. Once we have such a strategy we can give a detailed algorithm, and, if we desire, implement it in some executable language.

In contrast, consider how we approach the problem using logic programming. Here we write two clauses describing the relationship between the three lists, one to cover the special case where the first list is empty and the other to cover the general case. We thus get the following clauses:

append([],List,List).
append([First | Rest],List,[First | NewRest])
← append(Rest,List,NewRest).

The first clause simply says that appending a list to the empty list will result in just that list. The second clause says that the first term of the output list will be the first term of the first list and the rest of the output list will be the rest of the first list appended to the second list. If we execute this using a suitable resolution strategy it will behave in the way suggested in the preceding paragraph.

Suppose we invoke the program with the goal append([a,b],[c,d],L). The first call to append will call the second clause with the unifier

{First: = a, Rest: = [b], List: = [c,d]}

and call append([b],[c,d],New Rest]). This again matches the second clause with unifier

{First': = b, Rest': = [], List': = [c,d]}

and call append([],[c,d],New Rest'). This time the first clause will match with unifier

{List": = [c,d], NewRest': = [c,d]}.

This will result in NewRest becoming instantiated to [First',c,d], i.e.[b,c,d], and hence the L of the original query to [First,b,c,d], i.e. [a,b,c,d], which is the desired output. Thus we can see that when the program is executed the effect is to follow the algorithm that was described above. The point is that the algorithm is not described explicitly, as would be necessary in an imperative language: the program describes only the relationships involved, and the control from the interpreter produces the algorithm.

As a second example, consider the problem of removing a given term from a list. Informally we could produce an algorithm along the following lines. Consider each element of the list in turn. If the element is not the one we wish to remove, push it onto a stack. If it is, throw it away and pop the elements on the stack and push them onto what remained of the list. Now consider how we would describe the relationship between the two inputs and the output. Again we will have two cases, one where the required element is the first term in the list, and the other where it is not. In the first case, the required output is simply the rest of the list. In the second, the first element of the output list will be the first element of the input list, and the rest of the output list will be the result of removing the required element from the input list. Thus

remove(Element,[Element | RestOfList],RestOfList).
remove(Element,[Other | RestOfList],[Other | RestOfOutput])
 ← remove(Element,RestOfList,RestOfOutput).

As in the case of append, we can rely on resolution and unification to produce the algorithm for us given only the above declarative specification of the relationships involved.

7.3.2. Reversibility of predicates

There is, however, an important consequence of just defining the declarative relationship between the inputs and the outputs in this way, and not stipulating any control, namely that it means that these definitions are entirely neutral as to which of the terms are input and which are output. In the append example we had assumed that we would have two input lists and output the result of appending one to the other. But we do not need to rely on this assumption: we could equally well have a list as input and output pairs of lists which could be appended to give that list. Thus suppose we invoked append with only the third argument instantiated, say

append(First,Second,[a,b,c]).

Executing the above using the resolution strategy employed by PROLOG this would successively instantiate First to [] and Second to [a,b,c]; First to [a] and Second to [b,c]; First to [a,b] and Second to [c]; and, finally First to [a,b,c] and Second to []. The point here is that the logic of the relationships involved in the two problems, to append two lists to give a third and to discover the sublists of a list, is the same; thus the logic program is the same, and the detailed method of enforcing the relationship can be sorted out by the interpreter at run-time. In contrast, of course, if we are forced to specify the algorithm, we need to incorporate control as well as logic, and so must write different algorithms for the two problems.

The same neutrality as to input and output can be seen with the remove relation defined above. Here we can invoke the program with any one, any pair, or even all of the arguments uninstantiated and get sensible results. Thus if we take as goal remove(a,List,[b,c]), we will get a succession of answers corresponding to the different lists from which a could be removed to give [b,c], namely [a,b,c], [b,a,c], and [b,c,a]. Alternatively, if we left the first argument uninstantiated, and so used as goal remove(X,[a,b,c],[b,c]), we would get the answer that X is a. Leaving the first two uninstantiated, remove(X,List,[a,c]), we get the answers [X,a,c], [a,X,c], [a,c,X], indicating that whilst we cannot know what element has been removed, something must have been, and that it could have been removed from any of three positions. Thus if we have a family of problems which turn on the same logical relationships, but which apply different inputs to that logic and so generate a variety of outputs, we can use a single problem description, rather than needing to write a new algorithm for each input/output combination.

7.3.3. Logic as a representation paradigm

Having briefly introduced the central ideas of logic programming, we can now return to the idea of logic as a knowledge representation paradigm within the terms of AI. Hayes, it will be recalled from Section 7.1, decided to use first-order predicate calculus for his representation language for non-computational common-sense theories of the world because it was as expressive as competing theories, as metaphysically adequate as other theories, and because of its evident virtues in terms of lack of ambiguity. The properties of logic, and the techniques for representing facts and relationships are well established by the work done in formal logic. If we want a computationally tractable theory, then given the present state of the art we need to restrict ourselves to the Horn Clause subset of predicate calculus. This loss of expressiveness buys us computational adequacy, and also, if we accept the views of the proponents of logic programming, a good deal of heuristic adequacy, in that the essential feature of logic programming is that the knowledge required to solve a problem can be represented declaratively by means of a logic program. Section 10.2 will discuss whether the price paid for these computational advantages is too high when viewed as a representation paradigm.

References

1. Hayes, P.J., "The second naive physics manifesto", in *Formal Theories of the Common Sense World* (ed. J. Hobbs and R.C. Moore), Ablex, New Jersey, 1985.
2. Robinson, J.A., "A machine-oriented logic based on the principle of resolution", *Journal of the ACM*, **12**, (1965), pp. 23–41.
3. Bledsoe, A., "Non-resolution theorem proving", *Artificial Intelligence*, **9**, (1977), pp. 1–35.
4. Wos, L., Overbeck, R., Lusk, E. and Boyle, J., *Automated Reasoning: Introduction and Applications*, Prentice Hall, Englewood Cliffs, N.J., 1984.
5. Hogger, C.J., *Introduction to Logic Programming*, Academic Press, London, 1984, p. 40ff.
6. Kowalski, R.A., *Logic for Problem Solving*, North Holland, Amsterdam, 1977, p. 125.

8
PROLOG

The discussion in the preceding three chapters has described three knowledge representation paradigms. In all three cases, however, the ideas of these paradigms have become strongly associated with a programming language, so that using these programming languages can become identified with using the corresponding knowledge representation paradigm. Thus OPS5 is based on the production-rule paradigm, object-orientated languages such as Smalltalk on the structured object paradigm, and PROLOG on the predicate logic paradigm. It is worth considering the relation between the underlying paradigms and the programming languages, since the comparison shows up some interesting differences between the representation paradigms and the programming languages which indicate corresponding differences in the activities of knowledge representation and programming. In this chapter, therefore, I shall consider one such language, PROLOG, in some detail. I have chosen PROLOG for this detailed consideration firstly because it is widely taught and more widely available that either OPS5 or Smalltalk, and secondly because the well defined semantics of logic means that the predicate logic paradigm is somewhat more sharply defined than the others, so that the differences between the paradigm and the language are clearer.

PROLOG is a practical programming language based on the ideas of logic programming. Essentially PROLOG allows the writing of programs as sets of Horn Clauses, and will execute these programs by means of resolution, applied in a top-down manner, so that the effect is like a goal-driven system, using a depth-first search strategy. Additionally, PROLOG selects clauses by trying them in the order they appear in the database, and evaluates the clauses in the body of the Horn Clause left to right. These choices as to a control structure are not in conflict with the spirit of logic programming, and represent a good set of choices from the computational point of view. They do, however, have features which can be exploited so as to give PROLOG programs a behaviour different from that which we might expect from a purely logical standpoint. PROLOG, then, is worth attention both as a concrete example of logic programming, and because it illustrates some of the compromises that need to be made to produce a practical language.

8.1. Features of PROLOG

This brief chapter cannot be seen as a general introduction to the PROLOG language, but only to those features which are needed for a consideration of the language as it is relevant to knowledge representation issues. Readers needing an introduction to PROLOG should consult one of the many books serving this purpose such as Clocksin and Mellish [1] or Crookes [2]. The standard syntax of PROLOG is that known as Edinburgh PROLOG, as it first appeared in an implementation development at Edinburgh University. The following conventions apply:

1. ← is written as :-
2. & is written as ,
3. ∨ is written as ; (note that ∨ is not strictly required, but is provided for convenience in some cases)
4. Constants and predicate names begin with a lower case letter
5. Variables begin with an upper case letter
6. Parameters to predicates are enclosed in brackets and separated by commas
7. Lists of terms are written in square brackets and the terms separated by commas
8. The symbol | is used to separate the tail of a list: [A, B | C] represents a list of at least two terms, C representing a list of 0 or more terms
9. All clauses end with a period

So the following represents a PROLOG program dealing with a set of family relationships:

```
father(X,Y):- male(X), parent(X,Y).
mother(X,Y):- female(X), parent(X,Y).
grandparent(X,Y):- parent(X,Z), parent(Z,Y).
parent(bess,liz).
parent(george,liz).
parent(liz,anne).
parent(phil,anne).
parent(liz,chas).
parent(phil,chas).
parent(chas,will).
parent(di,will).
parent(anne,pete).
```

sex(george,male).
sex(chas,male).
sex(phil,male).
sex(will,male).
sex(bess,female).
sex(liz,female).
sex(anne,female).
sex(di,female).

We execute this program by typing in a goal to the PROLOG interpreter. So, if, for example, we type father(phil, chas), we will receive the answer "yes", indicating that the goal is a consequence of the axioms; whereas the goal father(chas,phil) will receive the answer "no", indicating that this goal is not a consequence of the axioms. We can use variable in goals, so, for example the goal father(X,chas), would receive the answer "X = phil", indicating that the goal is a consequence of the axioms with phil substituted for X, i.e. father(phil,chas) is a model for father(X,chas). If there are multiple substitutions for the variables in the goals, the first solution will be found and then the user may request further solutions by typing a ";". Thus grandparent(X,will), will first return "X = liz", and, if a further solution is requested, return "X = phil". If a further solution is requested it will next return "no", indicating that there are no more models for the goal. Finally we can give conjunctive goals such as "grandparent(X,will), father(X,Y)" (so that we can find the grandfather of will); this will return X = phil Y = chas"; X = liz of course satisfying only the first conjunct.

As can be seen from this brief description, PROLOG provides a straightforward translation of Horn Clauses, and a means of executing them so as to determine whether a given clause is a consequence or not, and what substitutions are necessary for a clause containing variables to be a consequence, and so is a valuable tool for exploring the claims and principles of logic programming.

If we look at the way in which PROLOG is used we find that the style of use can usefully be seen in three ways: as an implementation of logic programming, as a deductive database, and as an "AI" programming language. In practice, a largish application may exhibit all these styles of use, but they have a somewhat different flavour in the various cases, and it is worth making some remarks about each in turn.

8.2. PROLOG for logic programming

In this style emphasis is given to the ability to specify a program as a set of Horn Clauses, and to execute this specification using PROLOG's control strategy. As an example of this we can consider the quicksort algorithm. The strategy used by quicksort is to split a list into two sublists according to whether they should come before or after the first term in the original list. This procedure is then applied to the sublists until the resulting sublists become empty, when the various sublists are reassembled to give a sorted version of the original list. An example may clarify:

> Suppose we wish to sort the list [c,b,g,e,a]:
> First we split it into two lists, one containing elements less than c, and one containing elements greater than or equal to c, thus: [b,a] and [g,e].
> These lists are split on b and g respectively to give [a] and [] and [e] and [].
> Splitting these lists gives four empty lists.
> Working back we first get the lists [a,b] and [e,g] and by going back one step further [a,b,c,e,g], which is the original list sorted.

Let us the express this algorithm in Horn Clauses. First we will need a predicate *sort* with two parameters, the first for the list to be sorted and the second for the sorted list. One clause will express the fact that the empty list is sorted:

> sort ([],[]).

If the list is not empty, however, we must split it into sublists, sort those sublists and append the results. Thus

> sort([Head | Tail],SortedList):-
> split(Tail,Head,LessThanHead,GreaterThanHead),
>   sort(LessThanHead,SortedLess),
>   sort(GreaterThanHead,SortedGreater),
>   append(SortedLess,[Head | SortedGreater],
>   SortedList).

Now we need to define *split*. Again this has a base case, where the list is empty, and so will split to empty sublists:

> split([],A,[],[]).

A non-empty list will require two clauses, one for the case where the first term is less than the comparison term, where it must be put in the list built in the third parameter, and the other where it is greater and must be put in the list built in the fourth parameter.

split([Head | Tail],Term,[Head | LessThanTerm],
GreaterThanTerm):-
    Head = <Term,
    split(Tail,Term,LessThanTerm,GreaterThanTerm).
split([Head | Tail],Term,LessThanTerm,
[Head | GreaterThanTerm]):-
    Head > Term,
    split(Tail,Term,LessThanTerm,GreaterThanTerm).

Lastly we need a definition of append, as discussed in Section 7.3.1:

append([],List,List).
append([Head | Tail],List,[Head | Whole Tail]):-
    append(Tail,List,WholeTail).

This is no more that a set of Horn Clauses giving a formal description of the logic of the quicksort algorithm, and it is immediately executable using PROLOG. It is, given a rudimentary knowledge of PROLOG, no more difficult to understand than the verbal description given above, and has the advantage of being formally precise and the further advantage of being able to generate examples of its operation, by being executed. The program is completely declarative and does not depend on any of the features of control strategy imposed by the language. It would be possible to exploit knowledge of that control strategy to produce a (slightly) more efficient program by omitting the comparison from the third clause for split. This is possible because we know that the second clause will be tested before the third so that if the third clause is reached, Head must be greater than Term. To do this would be good programming but bad representation since the third clause of split would no longer be a true statement, but would rely on the context in which it will be invoked to give the correct behaviour.

Thus we can use PROLOG as a vehicle to do logic programming putting the dictum of algorithm = logic + control into practice. This enables us to experiment with features of logic programming, such as the efficacy of logic as a specification language, program verification and transformation and the like. From the standpoint of representation, such a style allows us to express problem-solving knowledge by giving a description of the relationships between the initial state and the final state that constitutes a solution.

8.3. PROLOG as a deductive database

Another use of a logic programming language such as PROLOG, is to build a deductive database. Such a program will look a little like the family information example given in Section 8.1. Relational databases are now widely accepted as the best database paradigm for databases, and the tables of a relational database have an obvious correspondence to the relations of a logic program expressed as facts. Thus:

person	mother	grandmother
chas	liz	bess
anne	liz	bess
will	di	liz
pete	anne	liz

This relational database represents an extensional description of the relations, since it simply lists all the participants in the various relations. As such it involves a degree of redundancy. In the example above, for example the grandmothers of a person are explicitly recorded, even though this information could be deduced from that in the other columns. But, as we saw from the example program above, we can give an intensional description (that is one which specifies a condition that represents necessary and sufficient conditions for fulfillment) of this relation in PROLOG by giving a clause which describes the relation in terms of some primitive relations. This means that we do not need to store the redundant information, but need store only the primitive relations and intensional descriptions of the other relations required.

Of course, in a conventional relational database we could store only the primitive relations and require the user to retrieve the other information by phrasing his query in terms of those primitives. Whilst this will achieve the same result, the effect is that the person querying the database must supply the appropriate intensional description afresh each time. Quite apart from the unnecessary typing, this means that he needs to know which relations are primitive, although this might not be clear. In the example parent and sex were taken as primitive. But we could have taken father and mother as primitives and defined parent and sex in terms of them:

```
parent(X,Y):-father(X,Y).
parent(X,Y):-mother(X,Y).
sex(X,male):-father(X,Y).
sex(X,female):-mother(X,Y).
```

This would probably have been a less sensible choice, since we would be unable to determine the sex of non-parents, but the lesson that it will not always be clear which relations the designer will have chosen to make primitive is no less valid. The advantages of a deductive database are that it eliminates redundancy whilst relieving the user of the need to compose afresh queries regarding non-primitive relations each time he needs the information, and by making the difference between primitives and non-primitives transparent.

Considered from the standpoint of knowledge representation, the deductive database style gives us a very good starting point. Thinking back to the semantic net representations described in Section 6.1, we will recall that the idea was to represent a number of binary relations, and to implement pointer-following procedures to allow extra information to be deduced from this representation. The deductive database offers a much cleaner model: the database of primitives can represent all the binary relations of the semantic net and the intensional descriptions of non-primitive relations can take the place of the pointer-following procedures, whilst remaining within the same overall formal framework, since both the intensional descriptions and the facts are expressed in the Horn Clause subset of predicate logic. In addition, we are not restricted to binary relations, and so where it is natural to represent a relation such as giving a relation between three objects we can do so and do not need to invent pseudo-objects such as the giving event, as in Section 5.1.1, nor to proliferate relations unnecessarily.

In short, using PROLOG to implement a deductive database enables us to represent a good deal of information and to use deduction as a means of supplying extra information, using intensional descriptions of the non-primitive relations, from the given information. For this reason the model has been popular in AI, and this style of programming will normally form part of any PROLOG AI application. There are, however, doubters: these are people who do not accept that deduction is a good model of human reasoning, and therefore claim that a deductive database approach cannot reflect accurately the use people make of information. Central to such doubts are non-monotonic and inexact reasoning, which we will consider in Sections 10.3 and 10.4.

8.4. Non-logical features of PROLOG

As PROLOG provided a means of demonstrating how useful the underlying ideas of logic programming and deductive databases were, use of the language started to increase, and the desire to make it into a fully fledged programming language was stimulated. It was found, however, that the original conception had a number of frustrating deficiencies when viewed in this light, and it was found necessary to build into PROLOG several features which deviate from the logic programming paradigm, and which can therefore be conveniently termed *non-logical*, or *extra-logical* features. These fall into three classes: features to facilitate input–output, features to affect the database in the course of program execution, and features to affect control. These features are worth considering in some detail because they help to point up some of the differences in the requirements of a programming language from those of a knowledge representation scheme.

8.4.1. Features for input and output

First there are the input-output facilities. Pure logical PROLOG permits the user only to input a goal, and will output only "yes", "no", or the bindings (instantiations) of variables in the goal. Whilst this might be enough for the logic programmer, practical application building demands that the program be capable of accepting input from the user and displaying the output in a rather more formatted and user-friendly form. Additionally, it may be necessary to display intermediate results to the user. Consider a program which takes a number from the user, and squares it, and then performs another calculation if the user desires. All that will be put out to the user is a blunt "yes" or "no" at the end of the execution, since the original query contained no variable whose bindings could be reported. Therefore PROLOG is equipped with a predicate *read* which takes one parameter which is bound to the user's keyboard input, and a predicate *write* which also takes a single argument which is printed out to the screen. Both of these predicates will always evaluate to true, and so will never make any contribution to the truth conditions of the predicate in the head of the clause. They are used solely for their *side-effects*; that they will cause something to happen, rather than for any logical purpose. The use of these two predicates is illustrated by this PROLOG program to perform the above task;

> calculate__squares:-write("Number to square:"),
> read(N),
> square(N,Nsquared),
> write(Nsquared),nl,
> write("Another number?"),
> read(no);calculate__squares.

A suitable definition of square is assumed. This short program also contains two other features of interest: the "nl" predicate which is another extra-logical predicate which simply supplies a line-feed, and the use of ";" which is used to ensure that the program ends if the answer is no, and otherwise repeats itself. Normally we can obviate the need for disjunction by having two clauses which may share the same conditions. In a program with no non-logical features this will at worst mean that some goals are evaluated unnecessarily as in this simple program to determine whether a list contains either the element 0 or 1:

> contains__0__or__1(List):-member(X,List),X= =0.
> contains__0__or__1(List):-member(X,List),X= =1.

which could have been written using ";" as

> contains__0__or__1(List):-member(X,List),X = = 0;X = = 1.

saving some calls to member.

In the squares case, however, we could not avoid the disjuncion by writing

> calculate__squares:-write("Number to square:"),
> read(N),
> square(N,Nsquared),
> write(Nsquared),nl,
> write("Another number?"),
> read(no).
>
> calculate__squares:-write("Number to square:"),
> read(N),
> square(N,Nsquared),
> write(Nsquared),nl,
> write("Another number?"),
> read(yes),
> calculate__squares.

as this would involve doing all the read and writes again. We could avoid the repetition of the side-effects and the use of a disjunction by writing

> calculate__squares:-write("Number to square:"),
> read(N),
> square(N,Nsquared),
> write(Nsquared),nl,
> write("Another number?"),
> read(no).
> calculate__squares:-calculate__squares.

which would give the correct behaviour since the second clause would be reached only if the answer was other than "no", but that would again be non-logical since the second clause, read declaratively, is untrue. This dilemma, the tension between writing declaratively and achieving the correct behaviour, illustrates the potential difficulties that come from deviating from the logic programming style and using predicates only for their side-effects rather than for their contributions to the truth of the head of the clause when there is some possibility of a need to backtrack. Notice too how procedural in nature the program is, listing one action after another to be performed rather than providing definitions of relationships, and how the behaviour is crucially dependent on the order in which the clauses appear.

Despite these disadvantages, the use of predicates to effect input–output is an evil which it is necessary to live with if PROLOG is to be a fully fledged

132 Knowledge Representation

programming language. Moreover, since they will always evaluate to true, they can simply be ignored in giving a declarative interpretation to a program, since they cannot affect the success or failure of a goal. In the example above we can read calculate_squares declaratively as saying that Nsquared is the square of N, which is true provided we do not worry where N comes from. Other predicates of this sort are provided to enable files to be loaded and saved, and for control characters and escape sequences to be issued.

8.4.2. Features to affect the database

The second group of extra-logical predicates concern affecting the database, either by adding or retracting clauses during program execution. Again there can be no logical reason for wanting to do this: if a thing is provable from the original database it can be proved again, and so it is unnecessary to assert it, and if it cannot, asserting it means that a consequence is provable from the altered database which was not provable from the original database, which means that we are not proving only the consequences of the original database, which is what we expect from a logic program. Retracting clauses means that things cease to be provable, which goes against the monotonicity of classical logic. Despite this, it is often convenient to be able to do these things, either for efficiency, so that we need not spend time re-proving a result that has already been proved, or to enable the implementation of certain forms of program. Therefore three predicates are provided, each of which take a clause as their parameter: *asserta*, which adds the clause to the beginning of the procedure for that clause, *assertz*, which adds the clause to the end of the procedure for that clause, and *retract*, which removes the clause from the database. A program illustrating a way in which these predicates might be used will be given after we have discussed the third type of extra-logical feature, which is also used by the program.

8.4.3. Features to affect control

The third type of extra-logical predicate affects the control of the program. Normally a program will test clauses in turn until one fails, at which point it will backtrack to the most recent choice point and seek an alternative means of satisfying that goal. If no clauses fail, no backtracking will occur. There are occasions, however, when the programmer wants to control whether or not backtracking will take place. PROLOG enables this by supplying two predicates, the *cut*, written "!" in Edinburgh syntax, which prevents backtracking beyond the point at which it appears in the clause, and *fail*, which always fails, and so can be used to force immediate backtracking.

To see how these might be used consider the following program:

    old(X):-age(Y,A),A > 65.
    age(jo,30).
    age(bill,76).
    . . .
    age(zeke,98).

If we use this program to pose the query old(jo), the program will first bind A to 30, and so fail. It will then attempt to satisfy age(jo,A), and so need to search through all the remaining clauses for age in the database. This may well be a lengthy process (lengthier still if the predicate concerned is not primitive), and quite unnecessary since the programmer is aware that everyone only has one age, so that the first failure is enough to immediately fail old(jo). Therefore once we have found an age for a person we do not wish to backtrack. So we amend the program by placing a cut after age:

    old(X):-age(X,A),!,A > 65.

Now when A>65 fails, the cut will prevent any attempt to re-satisfy age(X,A), and so old(X), will fail immediately. Note, however, that doing this will prevent a general use of old(X). For, suppose we wished to use it to generate a list of all the old people in the database: now the cut will mean that the test is applied only to the first item for age in the database. Thus if we use a cut, we need to be sure that we will never want the backtracking; in the case above, for example, we must be confident that we will only call old(X) with X instantiated.

Use of the cut can also distort the sense of the clause in our databases. All people born in the normal manner have navels, but Adam and Eve, being not born of woman, did not. We can record this information by the following clauses:

    has__Navel(adam,no):-!.
    has__Navel(eve,no):-!.
    has__Navel(X,yes):-person(X).

This program will behave sensibly, but only because the use of cut means that the third clause will not be reached if the first parameter is Adam or Eve. Strictly the third clause says, falsely, that all persons have a navel, including Adam and Eve. A correct declarative rendering of the information would be

    has__Navel(X,yes):-person(X),X = / = adam,
            X = / = eve.
    has__Navel(adam,no).
    has__Navel(eve,no).

Now only one of the clauses can be satisfied for a given value of X, and the behaviour is in line with the declarative reading. Opinions about the use of cut vary; some say that it is pernicious and should always be avoided, since it tends to encourage a lazy programming style, and is often used as a substitute for thinking precisely about the correct definition of the predicate. Others regard it is an essential feature of PROLOG as a practical language, both to avoid inefficiencies, and because the use of cut can improve the readability of the program. This is clearly a matter of taste, therefore: from a representation standpoint we should note only that its availability and use tends to promote the representation of procedural information, and to allow non-declarative clauses to intrude into our knowledge base.

The other control-affecting predicate is *fail*. This is often used to repeat side-effects. As an example of the this use consider:

> printAll(P):-call(P)write(P),nl,fail.
> printAll(P).

This predicate will, when supplied with an argument, print all the matching clauses present in the database. Thus if we call printAll(age(X,Y)), this will list all the clauses for age in our database, with the variables appropriately instantiated: age(jo,30), etc. Without the fail, of course, only one such clause would be printed; in order to find the rest we need to force backtracking. The second clause is there to ensure that the procedure as a whole succeeds. After the last matching clause has been found, and backtracking will fail to re-satisfy write(P), the second clause will simply succeed without further ado.

Fail can also be used in conjunction with cut if we want to specify some conditions which will ensure that a goal does not succeed. Suppose we have a predicate canFly(X), which is supposed to succeed only if X can fly. We therefore do not want it to succeed if X is a penguin or an ostrich, which we can achieve with

> canFly(X):-is-a(ostrich,X),!,fail.
> canFly(X):-is-a(penguin,X),!,fail.
> canFly(X):-is-a(Y,X),is-a(bird,Y).

In this instance we are doing something very similar to the Adam and Eve case above, except here we are using fail to mark the failure of the general rule in certain specified cases rather than a second parameter.

8.4.4. Implementation of negation as failure

One other use of cut and fail is worth remarking on. PROLOG comes supplied with a built-in predicate *not* which takes a clause as its parameter, and which succeeds if that clause fails (which parallels another built-in predicate *call* which succeeds if the clause given succeeds). This could be implemented were it not provided by

> not(P):-call(P),!,fail.
> not(P).

If P succeeds from the database call(P) in the first clause will succeed, and fail will ensure that not(P) fails, since the cut prevents the second clause being tried. If call(P) fails, of course, the cut is not reached and so backtracking will ensure that second clause is reached, and this will immediately succeed not(P).

8.4.5. Example of use of non-logical features

As a final example of the non-logical features, consider the following program. Again it shows the implementation of a built-in predicate, this time "findall". This program is intended to be self-explanatory, and also illustrates the use of comments (lines preceded by a "%"):

> % findall is a built in predicate. ownFindAll shows how it
> % could be implemented if not built in.
> % 1st argument is a variable which also appears in a Goal
> % given
> % as 2nd argument. 3rd argument will become instantiated to
> % a list of all terms which would satisfy the goal when
> % substituted
> % for the appropriate variable. Shows use of assert, retract,
> % cut.
>
> ownFindAll(X,Goal,Xlist):-call(Goal),assertz(queue(X)),fail.
> ownFindAll(X,Goal,Xlist):-assertz(queue(bottom)),
> collect(Xlist).
>
> collect(L):-queue(X),X \ = = bottom,retract(queue(X)),!,
>   collect(Rest),L = [X | Rest].
> collect([]):- retract(queue(bottom)).

Note particularly the use of fail to force backtracking so that all matching clauses are found, the use of cut to inhibit unwanted backtracking in the first clause for collect, and the use of assertions in the database to hold the intermediate results. Notice too how far such a program is from the declarative notion of logic programming: the thinking here is entirely in terms of how the behaviour can be achieved rather than in terms of the logic of the problem. Not that this makes it a bad program: in truth some programming tasks are inherently procedural, and the non-logical extensions are needed for PROLOG to be a practical programming language so that it can cope with such tasks.

I have dealt at some length with the non-logical features of PROLOG because it helps to point up the difference between a PROLOG program and a set of Horn Clauses considered as a representation of knowledge. The non-logical features represent no declarative knowledge, although they are often required to achieve the right behaviour for the program. In a sense such programs do represent knowledge, albeit of a procedural kind, and the need to extend the logical representation to allow this knowledge to be expressed points at some expressive limitation of the part of first-order predicate calculus. Sometimes, as in the ownFindAll program, the knowledge may be entirely procedural, in other cases exemplified by the use of cut, extra knowledge as to how the representation is to be used is required since the standard application of the proof strategy leads to the wrong behaviour, as when we wanted to represent the fact that there was only one model for age(jo,A). For these reasons we need when considering an application built in PROLOG to keep in mind the distinction between three things: logic as a representation of factual knowledge, best shown by the deductive database style, logic as a representation of problem-solving knowledge where the addition of control yields an algorithm, and those aspects of the application which are entirely procedural in nature.

8.4.6. Arithmetic in PROLOG

Before leaving this topic of non-logical features, PROLOG's treatment of arithmetic should be mentioned. The standard way of doing arithmetic is to use "is". This is done by writing a term followed by "is" followed by an arithmetic expression. For example,

>V is 3 + 4 will succeed with V bound to 7
>10 is 5 * 2 will succeed
>and 3 is 6 − 4 will fail.

The use of this is restricted, however, because an error will be given if there is an unbound variable of the right-hand side. So

>7 is V + 3
>will succeed only if V is bound to 4 when the clause is evaluated.

This means that we cannot use "is" to define an arithmetic relationship between variables declaratively: we must know the context in which the clause will be evaluated so that we can be certain that all the variables on the right-hand side will be bound. This is sometimes a restriction, and means that if we need to express such a relationship we must define our own predicates. For example,

> sum(A,B,C):-integer(A),integer(B),C is A + B.
> sum(A,B,C):-integer(A),integer(C),B is C − A.
> sum(A,B,C):-integer(B),integer(C),A is C − B.

will declaratively express the relationship between three integers, and will work as long as at most one of the variables is unbound. This is still very inflexible, and still requires a degree of knowledge as to the context in which it will be used.

8.5. PROLOG as an AI programming language

The development of PROLOG by the inclusion of extra-logical features parallels in some ways the developments of LISP. That language can be considered as a functional language, but the inclusion of many features to support procedural programming and programming for side-effects has led it rather to be considered simply as a programming language. It has been found very suitable for AI programming, particularly because of the flexibility of the list as a data structure, the scope it offers for writing programs that can be manipulated by other programs, and the ability to build the application incrementally. So closely associated with AI is LISP that an AI text book can state:

> In the study of Artificial Intelligence the reason for learning LISP is roughly that for learning French if you were going to France − it's the native language [3].

However, the features that made LISP popular for AI are also found in PROLOG, and that language has become, with wider availability and better implementations, a strong competitor to LISP for AI purposes. In addition, PROLOG offers for free three additional features: an indexed database containing clauses which can represent either procedures or data; a pattern-matching facility in the form of its unification algorithm; and a built-in search strategy giving depth-first goal-driven search and backtracking. Although these features, especially the last, may sometimes prove something of a mixed blessing, they do enable several of the things needed to write AI programs to be done with ease and efficiency.

People who program in this style need not be concerned with the principles of logic programming, and are not necessarily committed to logic as a representation paradigm, or deduction as a model of human reasoning. For them PROLOG can be regarded simply as a tool for implementing their favoured representation and reasoning styles. It is therefore important to distinguish between an application which is merely built in PROLOG and one which makes these commitments to logic, otherwise false expectations may be encouraged.

8.6. Summary

Thus although PROLOG can be seen as a realisation of the predicate logic approach to knowledge representation and the logic programming approach, it is essential not to make the identification complete, since PROLOG is at once both more and less than an instantiation of this approach. It is less because it implements only a subset of first-order predicate calculus, and so the ideal of representing knowledge in first-order predicate calculus and then running a program to manipulate this knowledge cannot always be met. Although the subset provided is very flexible and highly useful, it does mean that we must often distort or bend what we actually want to say to fit the restricted expressive power of the subset, which does have its limitations. It is more in the sense that it provides a general-purpose programming language well adapted to use in AI applications, and independent of the view taken of representation and reasoning. Care thus must be taken to distinguish the language from the philosophical approach to knowledge representation.

Exercises

8.1 Define the following terms as used in predicate logic, giving an example of each:

    literal
    clause
    Horn Clause
    well-formed-formula
    Skolem function
    modus ponens inference

Using predicates $F(x,y)$, $S(x,y)$ and $B(x,y)$ to represent that x is the father/sibling/brother, respectively, of y, and $M(x)$ to represent that x is male, write down predicates for the following axioms:

(1) All fathers are male.
(2) Two children with the same father are siblings.
(3) A brother is a male sibling.

Given also that

    F(john, harry)
    F(john, sam)
    F(sam, mary)

prove by resolution that

    harry has a brother.

Write brief comments to explain and justify the steps of your proof, and indicate where in your proof the brother of Harry can be identified.

[*Note*: A sibling of X is a brother or sister of X.]

8.2 (i) What is meant by the most general unifier of two clauses? What is the most general unifier of f([P,d | R]) and f([a,Q,c,d])?

(ii) In a PROLOG program, calling the procedure integers (I) will result in I being unified with an ordered list of the integers from 1 to 100 inclusive.

    (a) Write a PROLOG procedure for squares(S) which, when it is called, will result in S being unified with a list of the squares of the integers from 1 to 100 inclusive.

    (b) Write a PROLOG procedure for integers__and__squares(L) which, when called, will result in L being unified with a list of 100 pairs, each pair being an integer from 1 to 100 and its square.

In both (a) and (b) you must define any predicates used in the body of your procedures, other than integers(I).

8.3 (a) Give the unifiers for the following examples.
    (i) Parent predicates:
        p(U,U) and p(V,7)
    (ii) Parent predicates:
        q(f(Z), Z, f(7))   and   q(X,Y,f(Y))
    (iii) Goal:
        ?- ordered([4,6,5,7])
    Rule:
        ordered ([U,V | W]):- U < V, ordered ([V | W]).

(b) What new goal would be created by the application of the rule of the previous section to its goal?

8.4 (a) Give the most general unifiers of the following pairs of expressions:
    (i) append([a],[b,c,d],[a | Zs]) and append(X,Ys,[a,b,c,d])
    (ii) append(a,P,Q) and insert(R,[b,c],[r,s])
    (iii) append([b],[c,d],L) and append([X | Xs],Ys,[X | Zs])

(b) The procedure "select" is defined by the two clauses:
        select(X,[X | Xs],Xs).
        select(X,[Y | Ys],[Y | Zs]):-select(X,Ys,Zs).
    (i) What is the result of: ?-select(2,[2,3,2],P)?
    (ii) Explain exactly what unifications and goals are considered to arrive at this result.

(c) (i) Write a PROLOG procedure for first(X,Ys,Zs), which succeeds if Zs is the list obtained by removing the first occurrence of X from the list Ys. This procedure should fail if the list Ys does not contain X.

    (ii) Modify the procedure of the previous part to succeed if Ys is equal to Zs and Ys does not contain an X.

    (iii) What are the results of the procedure of part (ii) for the calls
        first(2,P,[3,4,2,5,6])
    and
        first(2,P,[3,4,5]) ?

8.5 Resolve the following pairs of clauses
    (i) P ← Q R
        Q ← S T
    (ii) P Q ← R S
        R ← T U

(iii) P ← Q R
      ← P
(iv) P ← Q R
     Q ←
(v) P Q ← R S
    R ← P

Why does the last example NOT resolve to Q ← S?

8.6 What are the non-logical features of PROLOG? Why do they go against the spirit of logic programming?

8.7 (i) Formalise the following argument in propositional calculus:

> Bob is either happy or Bob is working. Bob is not working. If Bob is not happy, then Bob is grumbling. So Bob is not grumbling.

Is the argument valid? Use a truth table to justify your answer.

(ii) Formalise the following sentences in predicate calculus:

> A wife is happy if her husband is happy.
> Bob is married to Jane.
> Alice is married to Phil.
> Bob and Alice are not both happy.
> If Bob is playing his trumpet, he is happy.

Can both Jane and Phil be happy? Can Phil be happy if Bob is playing his trumpet? Justify your answers.

(iii) Write a set of Horn Clauses based on your formalisation of the sentences in (ii) which could form the basis of a PROLOG program to determine whether or not someone is happy. Comment on any problems, and how you get round them.

(iv) Show how resolution can establish that Jane is happy from the program and the additional fact that Bob is playing his trumpet.

References

1. Clocksin, W.F., and Mellish, C.S., *Programming in Prolog*, Springer-Verlag, Berlin, 1981.
2. Crookes, D., *Introduction to Prolog Programming*, Prentice-Hall, Englewood Cliffs, N.J., 1987.
3. Charniak, E. and McDermott, D., *Introduction* to *Artificial Intelligence*, Addison-Wesley, Reading, Mass., 1985, p. 33.

9
Expert systems

This chapter will provide a discussion of expert systems, which currently represent the major effort in the commercial exploitation of AI. They are thus a fruitful area for exploring how knowledge representation ideas, particularly those relating to production rules, have been applied in practice, and have in themselves done much to extend the popularity of AI as a subject of study. In this chapter, therefore, we shall look at what expert systems are and why they are seen as applicable, and consider some illustrative examples which will also help flesh out the ideas sketched in Chapter 5, which dealt with production rules.

9.1. Why expert systems?

As we have seen in Chapters 1 and 4, initially AI saw its role as one of tackling hard general problems. Thus in games playing, chess, the intellectual game *par excellence*, was seen as an obvious choice (although the first program for which any degree of success was claimed was for the somewhat easier game of draughts [1]). The fact that language is a central concern of intelligent people led to experiments in the field of machine translation. Those of a mathematical bent saw the possibility of using computers to prove mathematical and logical theorems. And when AI was applied to more humble problem domains, of the level of IQ tests or colour-supplement brain teasers, it was felt right that a general problem solver should be developed. It is out of this last area that expert systems developed.

All of these enterprises met with only limited success. There were several factors which led to the failure of the general problem solvers. In order to achieve generality, the problems were couched in very abstract terms. Essentially, the strategy was to reduce the problem to a search problem, in which the initial state and the goal state could be characterised together with a set of operators which would transform the one into the other. Thus solving a problem was reduced to finding a path through the search space. As we have seen in Chapter 4, however, realistically sized problems have a search space which is too large to be tackled by the mechanistic exhaustive search strategies. Thus heuristics were required to guide and order the search. But finding heuristics of general applicability was difficult, if not impossible. The effect was that the more general the range of problems that could be tackled, the more unsatisfactory was the solution produced.

Secondly it was found that general problem solving employed a good deal of what might be termed "common-sense" knowledge. This is continually used by human problem solvers, and is taken by them as so obvious that it is hard to capture in a program, because it is so difficult to make explicit. For example, human reasoners will often argue from cases, identifying a set of possibilities and finding the solution by eliminating all but one of these candidates. This is a very difficult strategy to mechanise, in a general sense. So too are such facts, obvious to all, as that a person cannot be both male and female.

The key to all of this is, of course, that solving different problems tends to require different knowledge and different approaches. Whereas we can rely on human problem solvers to have a fairly extensive fund of common knowledge which they can bring to bear on a variety of problems, and which therefore we do not need to state in the problem specification, computer problem solvers need to be explicitly provided with all of this.

The obvious solution to these problems was to abandon the quest for a general problem solver and instead produce a system that would solve problems in a limited area. This would enable the knowledge representation used, the knowledge stored, and the problem-solving strategies employed to be tailored to that specific class of problems. Such a system would be a specialist rather than a generalist, an *expert*, for want of a better word. Hopefully, within that domain, such a system would be able to produce better and more efficient solutions to harder problems than was possible with the general-purpose problem solver. The term *expert system* often leads people to expect an unrealistically high level of performance; it is therefore worth bearing in mind that the term "expert" originally represented a limitation on the capabilities of the system, in that it is intended to imply firstly that it is applicable only to a narrow range of problems.

But the term is appropriate in another way. For fundamental to the restriction of the domain was that the system would be provided with a substantial body of knowledge in that domain, and that the problems solved in that domain should be substantial enough to compensate for the lack of breadth. This meant that the natural sort of domain to select for such a system was one which was associated with human experts: both because this indicated the problems were worth tackling, and because the existence of experts meant that they could be used as a source of the domain-specific knowledge that it was thought such a system would require. This was a particularly important move as far as exciting commercial interest went; it moved AI applications away from the unimpressive toy domains of the 8-puzzle and the blocks world into more impressive and potentially commercially viable areas such as medical diagnosis.

9.2. What is an expert?

From the above discussion we can see that the basic idea which led to expert systems was that problem solvers should be built which operated in a specific domain, which would be one associated with human experts. The idea was extended to suggest that these systems should mimic the behaviour of these experts in some sense. Thus the model on which the capabilities of such a system would be built would be that of a human expert. We need therefore to spend a little time looking at what we mean by a human "expert", and what aspects of his behaviour we wish to incorporate in our expert systems.

To begin with the negative aspects of expertise, we should note that the expertise of experts is invariably associated with a more or less narrow area in which they are expert. These specialities are typically quite specific; when we are ill we go to our general practitioner whose knowledge of medicine is broad, but who cannot claim to be an expert in any particular branch of medicine. Faced with a difficult problem he must refer his patient on to a consultant, and must use a different consultant for different problems. The same is true of lawyers, or computer scientists. The acquisition of expertise necessitates specialisation, and a restriction of interests to a particular field. Thus in designing an expert system we must first be careful to circumscribe the domain, perhaps even more narrowly than would be the case with human experts: an expert system to configure computers would not be expected to deal with computers in general, nor even the whole range of computers of a particular manufacturer, but rather with some particular model.

The selection of a domain is an essential part of designing an expert system, but so far we have no guidance as to what such a system would look like. For this we need to consider the way people interact with an expert. The key notion here is that experts are people we consult when we have a problem which falls within their area of expertise. It is this notion of a consultation that dictated the design of the first generation of expert systems.

The idea of a consultation provides a number of useful guidelines. Firstly, a consultation is initiated by a person with a problem. There is a specific question or problem to which he wants an answer. We would therefore expect the first stage of interaction with an expert system to be the user of the system posing a question.

These questions are not in the main general questions, but questions which relate to the particular circumstances of the person with the problem. We don't ask a doctor what causes peptic ulcers, but what has caused the peptic ulcer from which we are suffering at the present time. This means that once the expert has been confronted with the question he needs to elicit the facts pertaining to the individual concerned. The expert will therefore tend to ask his questioner a series of questions to get these facts. Although this series of questions might always begin in a similar way, the information supplied will tend to lead the expert to take a particular view, and subsequent questions will vary from case to case, as the expert attempts to confirm or disprove his hypotheses. The relevance and significance of questions will depend on earlier answers, and part at least of the expertise will consist of selecting the right questions. Thus stage two of the consultation is gathering information and using this information to hypothesise solutions to the problem and to determine what further information is required. At the end of this stage the expert will have arrived at a solution, and be satisfied that he has enough information to justify this solution.

At this point the expert will be able to tell us the answer to our question, or to advise us on our problem. So the stage of questions will end and an answer will be produced. This may satisfy the person consulting the expert, or further clarification and explanation of the answer may be sought. So the consultation may switch to a third stage in which the expert answers questions about his answer to the original question. Of course, some experts may be more forthcoming than others at this stage.

To summarise what is happening in a consultation with an expert: the expert has acquired from training and experience a large amount of generally applicable knowledge about his domain of expertise. The person consulting him has a particular problem relating to that domain. The purpose of the consultation is thus to bring the general knowledge to bear on the facts of a particular case, so as to provide a solution to the particular problem. The expert system modelled on this behaviour will similarly have represented within it the generally applicable knowledge, will be interactively supplied with the facts appropriate to a particular case, and will apply its knowledge to those facts to reach a solution.

9.3. What is an expert system?

We are now in a position to try to be a bit more precise about what we mean by an "expert system". The original idea was to develop systems which imitated experts in that they operated on a substantial problem in a limited domain, and in the consultative style in which they operated. As time has gone on, however, the term has been used to denote, in the words of Allan Newell writing in the foreword to [2], any system which is applied, has some vague connection with AI systems, and has pretensions of success. Such is the fate of terms that attain (if only briefly) a positive halo.

So indiscriminate has been the use of the term that expert systems have ceased to have quite this positive halo, and the more general term *knowledge-based systems* has tended to displace it as a term of recommendation for applied AI systems. This means that "expert" systems can be now used to identify a resonably coherent subset of knowledge-based systems which adhere relatively closely to the original conception. We should not, however, look for a tight definition, but rather a collection of features that we might expect such a system to have.

9.3.1. Expert systems have domains of expertise

Firstly an expert system should have a domain which is associated with human experts. Most of the classic expert systems satisfy this condition, but there is a distinctly grey area developing. For with the growth of shells, which will be discussed in Section 9.6, techniques originally developed within the expert systems context are being applied to applications which are too trivial to associate with expertise as usually understood. Popular usage probably would call these systems expert systems, in which case we are only left with the narrowness of the domain, without the depth.

9.3.2. Expert systems are interactive

A second feature is that an expert system should be interactive. The original expert system style was that of a consultant, and this remains the prevalent paradigm. Lately, however, different paradigms have developed, such as the critiquing expert system [3], where the user attempts to solve the problem himself and the system intervenes only when it considers the user is going wrong. Again it is a matter of taste whether this extension of the style of system is considered to take it out of the realm of expert systems or not. The features of interaction in a consultative expert system are essentially that it is able to ask questions of the user to obtain details of the problem situation, and that these questions are selected so as to be relevant to the particular problem, and that the system is capable of explaining its reasoning. The form of these explanations will be considered in Section 9.4.3.

9.3.3. Use of heuristics

Another feature we can look for in expert systems is that they tend to use heuristics rather than determinate rules. When expert systems builders attempt to get the domain knowledge from an expert they usually find that the information the expert comes up with tends not to be hard and fast, but rather the sort of thing which is usually applicable, but which may turn out to be false in exceptional circumstances. Moreover, experts will often disagree with one another as to the nature of this information and the extent of its applicability. Some people would go so far as to say that an expert systems domain necessarily contains this sort of defeasible information, since if it does not, problems are susceptible to a purely algorithmic solution, and expertise is unnecessary. Again the early systems did indeed make extensive use of defeasible heuristics, but currently more and more systems are built using expert systems techniques but applied to a body of certain knowledge, representing a down-grading of the original conception.

9.3.4. Uncertain and incomplete information

Next there is the ability of the system to handle uncertain or incomplete information. When the consultant asks questions, he may receive answers which are more or less vague or tentative, and sometimes the information may simply be unavailable. The designer of an expert system may wish to reflect these possibilities, and so design his system to cope with varying degrees of assurance in the user's answer, and to allow the user to admit ignorance in response to some questions. Yet again this was a very central feature in most of the original expert systems, but it appears to be far less important in most of the current systems to which the term is applied. We will return to the treatment of uncertainty in rule-based systems in Section 10.4.

9.3.5. Other possible features

In the formative stages of the development of expert systems there were a number of attributes that people felt that such systems might possess, but which no longer seem so central. Some argued that expert systems were necessarily rule-based: whilst it is true that this tended to be the dominant representation paradigm for early systems, there seems little reason to exclude other forms of representation. It is better to treat rule-based systems as another way of dividing the larger class of knowledge-based systems, independent of expert systems.

Another supposed feature was that expert systems should include not only object level knowledge of the domain of expertise, but also meta-knowledge which would allow reasoning about the limits of the object level knowledge and about appropriate strategies for using that knowledge, since there can be no doubt that human experts possess and employ such knowledge. In practice, few expert systems incorporated knowledge of this sort to any significant extent, and it seems that insisting on this requirement would withhold the title from the vast majority of existing systems. This is not therefore a useful distinction.

Experts, of course, converse with their questioners in natural language. We might therefore expect, or at least want, our expert systems to do the same. But a full natural language interface is still very much beyond the current state of the art, and so again there is little point in insisting on this feature. Expert systems have, however, paid a good deal of attention to the interface, and effort is taken to use techniques to make the interface seem as natural as possible. Many dialogues with expert systems do give a surface appearance of natural language, but this is largely apparent rather than real.

Finally, we may consider the topic of learning. Experts become expert through experience, and every consultation is an opportunity for them to learn more about their domain. This too was felt to be a desirable feature to incorporate in expert systems, so that they could start off with a knowledge base which would be automatically refined and adjusted in operation. Again this feature is no longer insisted upon because, desirable as it is, it is still very much a research issue, better pursued separately from the building of expert systems.

The situation is somewhat confused because expert systems started life as a research topic with high aspirations relating to imitating human experts across a wide range of features. Initial success with some of these features, however, led to such systems being taken up for commercial exploitation; most notably in the field of configuration of computer systems, to be discussed in Section 9.5.2, but now in several other fields as well, such as VAT auditing and advising on retirement pension entitlement, to be discussed in Section 9.9.2. Indeed, expert systems is one of the chief areas in which AI techniques can be used in a practical environment. Many of the original desiderata require, however, solutions to questions which are still open. This means that only some of the original aims can be realised in currently practical systems. To answer the question as to what an expert system is, therefore, requires something of a balancing act, steering a middle course between the original aspirations and the currently practical. The considerations which need to be balanced are those outlined above. It is my view that we can usefully see expert systems as an engineering problem whereby a certain set of techniques originally developed to explore AI research objectives (the modelling of expert behaviour) are applied to simpler, but commercially viable problems. I would therefore tend to apply a fairly liberal interpretation to the various criteria I have considered, although to do so involves the recognition that the nature of expert systems has changed somewhat during the move from the laboratory into the market place.

9.4. Basic expert systems components

In the last section I considered a number of features which expert systems may have to a greater or lesser extent. Despite the variation in these features found in realisations of expert systems, there is a remarkable consensus on the overall architecture of such systems. Almost all expert systems can be seen as comprising three distinct components, and it is worth making some remarks on each of these in turn.

9.4.1. Knowledge base

Central to any expert system will be a representation of the domain knowledge in a component termed the *knowledge base*. This is supposed to represent the domain knowledge elicited from the expert in some more or less declarative form, independently of how it is to be used to solve problems, and so free from details of control and implementation. Ideally it would just comprise a set of statements, in some suitable knowledge representation formalism, known to be true and describing the domain. In practice, however, the declarative ideal is rarely achieved, and some implicit control information gets incorporated within the knowledge base. For example, if production rules are our chosen formalism, we have already seen how the choice of conflict resolution strategy can alter the effect of the rules. Despite this qualification, the idea is fairly clear.

A major advantage of this approach is that the knowledge base can, given a sufficiently comprehensible representation, be understood by the experts who supplied the initial knowledge. Whilst it would be unreasonable to expect them to come to terms with a conventional encoding of their knowledge, because of the obfuscation that occurs with incorporation of control elements, they ought to able to understand the knowledge base of an expert system to the extent that they can assist in correcting, refining and validating it. The other major plus is that such a representation ought to be easier to construct; freed from the worries relating to control, the builder of the knowledge base can concentrate solely on representing the knowledge he has obtained.

9.4.2. Inference engine

The second component of the expert system will be the one which manipulates the knowledge base so as to solve the problems. This component is usually termed the *inference engine*, because it uses the knowledge in the knowledge base and the facts relating to the case under consideration to draw conclusions. The nature of the inference engine will depend on both the representation chosen for the knowledge base and the problem-solving strategy considered appropriate by the designer of the system. So for example, if the chosen representation is production rules, the inference engine will necessarily take the form of a rule interpreter, and different problem-solving strategies effected by decisions as to whether it is forward or backward chaining and the nature of the conflict resolution strategy employed.

9.4.3. User interface

The third component of expert systems is the *user interface*. All expert systems, whether consultative in style or not, are interactive, and so need a component to manage the interaction. The basic requirements on the user interface are to get requests for information from the user so that the session can be started, to ask relevant questions of the user and to convey the information on to other parts of the expert system, to present the advice or answers from the system to the user, and to provide the explanations required by the user.

Two types of explanation which are often provided in a rule-based system are explanations as to why a question is being asked, usually simply called a *Why?* explanation, and explanations as to how a conclusion was established, called a *How?* explanation. Given a backward chaining rule-based expert system, both of these explanations can be easily implemented from a consideration of the proof trace of the system. Suppose we have a rule base containing the following simple rules:

>R1 IF P & Q THEN R
>R2 IF S & T THEN P
>R3 IF R & U THEN V

and we wish to establish that V. The backward chaining system will match the RHS of R3 with this goal, and so add as subgoals R & U. R will match with R1 and P and Q will become subgoals. Now R2 will match and S and T will become subgoals. At this point no rules will match, and so the system will ask the user whether S is true. At this point the user may want to know why S is being asked about, since he may be unable to see any connection with V. If he requests a Why? explanation, he will receive something like this:

>I am trying to show that P:
>By R2 I can show P if I can show S & T

This is just a statement of the subgoal currently being pursued, and the rule being used in an effort to establish it. This may not satisfy the user and he may ask Why? again. This time the system will go one step back up its proof tree and offer the explanation:

> I am trying to show that R:
> By R1 I can show R if I can show P & Q

A final level of Why? explanation would step up the proof tree to reach the original goal:

> I am trying to show that V:
> By R3 I can show V if I can show R & U

At some point the user will become satisfied with the explanation, and answer the questions. In the example, once he has answered affirmatively to S and T, the system will ask about Q. Here the system would explain:

> I am trying to show that R:
> By R1 I can show R if I can show P & Q
> I can show that P.

Thus at any point in the proof, giving the position in the proof, the subgoal, the rule being used to show the subgoal, and any conditions in that rule already shown, will provide the reason for the question. Affirmative answers to Q and U will produce the conclusion V. At this point the user may ask how this conclusion was established, and receive an explanation something like

> By R3 I can show V if I can show R and U
> I can show that R
> I was told that U

Again this is a simple display of the rule used, and the means by which the conditions in the rule were established. Where the condition was shown by the system rather than supplied by the user, a further level of How? explanation may be sought, and a similar explanation giving the rule and the means by which the conditions were established: in the example, How R? would produce

> By R1 I can show R if I can show P and Q
> I can show that P
> I was told that Q

This kind of explanation can be sought until conditions which were all stated by the user to be true are reached.

Both these types of explanation are popular and widely used in rule-based expert systems. They are entirely sensible given the design of the system, although some people claim that they are difficult to understand, and not really the kind of explanation one would expect to get from a real expert. Much work is going on to discover what would constitute a satisfactory explanation, and how such explanations might be provided.

9.4.4. Some points about expert systems

Some key features presented by expert systems should be noted at this stage. Perhaps most importantly there is the central point that expert systems insist on the clear separation of the knowledge base, the information to be used to reach the solution, and the inference engine which manipulates that information. This is in stark contrast to the conventional program, which needs to provide a step-by-step algorithm for solving the problem, and which as a result inextricably entwines these two kinds of information. The separation is thus attractive, for many of the same reasons as the idea of declarative programming generally is attractive, in that it allows the system builder to concentrate on representation of knowledge and leave issues of control to be sorted out by the inference engine. This approach ought to lead to greater productivity, and gives the opportunity for several other important features.

A group of such features is related to the possibility of incremental development of such systems. Suppose we are building an expert system based on production rules. Provided we have a production rule interpreter to serve as an inference engine, we can produce an executable system with very few rules. This will of course not perform very well, but it does at least give some kind of feel for the way a full system will behave. This in turn offers the possibility of developing such a system through prototyping (or rapid prototyping as some call it). Because the inference engine is independent of the knowledge base, the initial system can run on a very limited knowledge base and give an indication of whether it is worth developing the ideas further or whether a radical re-think is necessary. At any stage of development of the system, an executable system will exist, and this speedy realisation of tangible results is a great help in reinforcing user acceptance.

A second sense of incremental development, which becomes increasingly evident as the knowledge base grows, is the possibility of using the current performance of the system to show where the system is deficient and where more development effort needs to be put. The original rules put into the system are likely to be only approximations: when the system is run, it will as a result give incorrect answers in some of the cases. The explanation facilities of the system can then be used to identify the rules which gave rise to these mistakes, and these rules can be examined to see why they misled the system. Typically there will be a condition applicable only in an exceptional circumstance which has been overlooked in the original version of the rule, and so this condition can be added, and the performance of the system enhanced so that it will give the correct answer to such cases in subsequent runs. Thus once a reasonably sized knowledge base has been produced, further development is a process of refining this knowledge base by a process of running the system, spotting any performance deficiencies, using the system to identify the source of these deficiencies, and correcting them.

The third way in which incremental development applies to expert systems is in the possibility of gradually extending the domain of the system. The target domain may be quite wide but susceptible of being seen as a number of related but relatively self-contained areas. If we are building a system to advise of the treatment of diseases, for example, we can begin by concentrating on one or two diseases and get the system working initially for these. When we have done that we can gradually widen the scope by incorporating further diseases, applying, of course, the lessons learnt from the initial diseases, until we cover the whole intended range. Because the inference engine does not change, and because the rules in the knowledge base should not have harmful interactions, this strategy of extension should be relatively painless. Compare this process with the extension of new facilities to a conventional program: anyone who has ever attempted to maintain conventional programs will know that in that case there are likely to be many problems, since the extensions will need to be incorporated in several scattered parts of the program and the lack of separation of control from knowledge means that harmful interactions between the existing code and the extensions will invariably arise.

A final key feature of expert systems is that the knowledge base should be intelligible to experts in the domain who are not themselves proficient with computers and computer programs. A key point about expert systems development is that the builder of the system, the knowledge engineer, should work very closely with the domain expert whose knowledge he is trying to encode. For this close relationship to be possible, the expert must be able to see the developing system in a form that he can understand, so that he can vouch for its correctness and suggest where it needs to be extended and refined. This is not at all possible in the case of a conventional language since the expert cannot be expected to be a skilled programmer. It is, however, expected in the case of an expert systems formalism, and this expectation has been fulfilled in some expert systems projects.

9.5. Early expert systems

Expert systems came to the attention of the general public through the appearance of a few high profile systems. These systems were associated with some extravagant claims, most of which they did not really live up to. None the less the ideas behind them were attractive and they spawned a host of imitators. In this section I shall mention a few of the best known and give some indication of the basic ideas they expressed. For details on such systems, see the specific references.

9.5.1. MYCIN

One of the best known of these early expert systems was MYCIN, a system intended to diagnose and suggest treatment for diseases of the blood, a development within the Heuristic Programming Project at Stanford University, begun in 1972. It was essentially an academic project, bringing together the contributions of a large group of researchers and Ph.D. students over a number of years. Although it was never actually used in practice to any great extent (claims vary as to the number of diagnoses it actually made, but it was certainly not used on any sort of regular basis), it has been enormously influential. This is partly because it embodied a number of important ideas, and partly because it was extremely well-documented in an excellent book by Buchanan and Shortliffe [4]. Essentially MYCIN was a straightforward production-rule system, offering a consultation style of interaction. Although the internal details are rather more complicated (unnecessarily so when viewed with hindsight), it can be seen as being a depth-first backward chaining expert system. Important ideas were the use of the rules to identify the relevant questions that needed to be put to the user, the close cooperation of experts to provide and validate the rules, attempts to use natural-language versions of the rules to enable easier communication with those experts, the use of certainty factors to handle incomplete and uncertain information and the provision of what became the classic explanation facilities of "How?" and "Why?". All of these things proved so persuasive that they are now almost the defining characteristics of an expert system.

9.5.2. XCON

Athough MYCIN proved highly influential in characterising expert systems and stimulating academic research, a second classic system, R1, later renamed XCON, did rather more to excite real interest in the technology in the commercial world. This is because this system was actually built for use in a commercial context, and has been used in that context to make an excellent return for its parent company, DEC, who make a number of minicomputers, most notably the VAX range. In addition, DEC had attempted to use conventional techniques to do this task before trying an expert system, and failed in these attempts. Thus the importance of XCON is that it demonstrated that expert systems could be a paying proposition, and could deal with tasks which had been found intractable using conventional techniques. A further point of influence was that the production-rule language developed as part of this project, OPS5, became a more or less standard language for production systems.

The task which XCON addresses is that of configuring DEC minicomputers. A minicomputer system is made up of a lot of separate components which are put together to meet the needs of the individual customer. The customer discussed his requirements with a salesman who prepared a list of components, making up the order for the system. The salesman did not, however, have the required expertise to configure these components into a functioning system, and the list would need to be sent back for an expert to perform this task. Often certain vital components would have been overlooked, and in addition the configuration was a time-consuming and tedious task which took the expert away from more profitable and enjoyable work. The idea of embodying this expertise in an expert system which could be used to check and configure orders by a person of less expertise, or even the salesman himself, was therefore an attractive business proposition, since it held out the prospect of freeing valuable expertise and increasing customer satisfaction by providing a quicker response and fewer errors.

The system produced was essentially a forward chaining production-rule expert system. Forward chaining was the obvious strategy since the initial list of components was given, and it would be an impossible task to start from the goal of achieving a legitimate configuration, since there were far too many possibilities. Two other interesting features were present in the system: the use of a different representation for the knowledge about the features of the various components and the knowledge of how to configure the systems, and the use of specificity as a conflict resolution principle.

The first of these features came from a realisation that much of the information regarding components was very static, and that this kind of information was consequently better represented in a static database than in production rules. For example the RK611* has class unibus module and type disk drive, and this could be represented as a rule such as "IF (? × component RK611*) THEN (? × class unibus-module) AND (? × type disk-drive) AND ..." However, it is more sensible (and efficient) to represent this information by means of entries in a database so that it can be accessed by look-up rather than being inferred afresh every time the system is run. This foreshadows the combination of structured objects and production rules which is used in many current systems.

The second point was again to do with efficiency. The configuration task had a fairly determinate order in which things should be done. First, the order needed to be checked, then the CPU configured, then the unibus modules configured, then the panelling, then a floor plan produced and finally the cabling designed. In order to impose this ordering of the task on the system, each of the rules has as its first condition a condition of the form "the most active current context is distribute mass bus devices", so that a check could be made as to which stage of configuration had been reached before considering the rule further. (*Context* is the term to describe the various stages of the configuration process.) This meant two things: first, that efficiency would be enhanced since only the rules applicable in the current context would be considered beyond the first condition, and second, that only rules relating to the current context could be fired, thus ensuring that the task was performed in the required order. Each context would also have a rule enabling the context to be changed, of the form "IF most active current context is this-context THEN most active current context is next-context". This rule will, because the conflict resolution strategy uses specificity, only fire when no other rules in the given context are applicable. The effect then is to change context when, and only when, all the tasks in a given context are complete.

The use of specificity had another beneficial implication. As the system was being developed and problems with the knowledge base were being discovered, corrections were often made by adding extra conditions to the rules. The error had arisen because the circumstances in which the rule was applicable had been underspecified. There was, however, no need to remove the old rule, which could still be of some use, since it would act as a default where the extra condition was not satisfied. This then provides a practical example of the use of general but defeasible rules discussed in Section 5.3.6. As a result, however, the current version of XCON is likely to contain some rules that are never fired at all, since the conditions can only be true where there is some more specific rule to apply. As a result of this kind of incremental development, the rule base has become somewhat unwieldy, and stands in some need of a rewrite.

To give some idea of the size of the system, the static database contains some 15 classes of unit, each with, on average some 8 attribute value pairs and about 50 different attributes spread across the different classes. The original version, current in January 1980, had about 750 production rules (enough to cope with the VAX 11/70), which had increased to about 3300 rules by 1984. On average the rules contained 6 conditions and 3 actions. The 3300 figure needs, however, to be considered along with the knowledge that some of the productions may be redundant and not reachable by any path open to the system for the reasons mentioned above.

XCON was a highly successful system and has been copied by several computer manufacturers faced with the same task, who have found similar success. Readers interested in more detail about XCON should see Ref. 5.

9.5.3. PROSPECTOR

The third of the early expert systems worthy of mention is PROSPECTOR, produced by SRI International in the late 1970s, which is interesting because it took a very different approach from the previous two, and was not based on production rules. The domain of application of this system was to suggest what mineral deposits were likely to found beneath a given piece of ground on the basis of field observations. Clearly this is a difficult task, but one with a very high payoff if it can be done successfully. The basis of the system was to record the probabilities of various mineral deposits being found given the evidence of the observations and to use Bayes' theorem to propagate these probabilities through the system. This way of handling uncertainty will be discussed further in Section 10.4. A good description of the system can be found in Ref. 6.

This system is worth mentioning here for the following reasons. First, it proposed to tackle a task of extreme difficulty, one that even the best available experts cannot perform with any degree of certainty. Its value would be realised only if it was capable of out-performing existing human experts, since the decision is so important to the company making it that the best available advice is well worth paying for. Second, it illustrates a very different approach to expert systems, one which does not rely on the declarative encoding of the knowledge used by an expert to solve the problem, but rather by manipulating probabilities thought to reflect some underlying truth about the domain. This approach, particularly in areas where the emphasis was on the uncertainty resident in the domain, enjoyed some popularity, but eventually has lost out to the declarative approach. The reasons for this are twofold: first, the limited success of the approach, but second, and more important, the difficulty of obtaining or estimating the probabilities required to form the basis of the system.

9.5.4. INTERNIST

A fourth expert system, providing an example of the use of structured objects in knowledge representation, is INTERNIST [7], designed to diagnose diseases in the field of internal medicine on the basis of observable symptoms. It holds its medical knowledge in a "disease tree", a hierarchical organisation of information relating to diseases in which the root node represents all diseases, non-terminal nodes disease areas (such as diseases of the lungs and diseases of the liver), and the terminal nodes specific diseases that can be diagnosed and treated. Given a set of symptoms, abduction is used to find the disease that best explains the symptoms. Knowledge of the disease can then be used to predict what other symptoms should also be present, and these can be used to confirm or disconfirm the abductive hypothesis. A detailed descripion of this process can be found in Ref. 8.

INTERNIST is a large system, the current version, called CADUCEUS, containing information on some 500 diseases. The size of this knowledge base meant that the exhaustive searching strategy used by MYCIN would have been impractical in this domain, and so the structure of the disease tree and the use of abduction provides a way of focusing attention of the relevant portions of the tree. It is an important example of how the structure inherent in a domain can be exploited in the problem-solving process.

9.5.5. Summary of early expert systems

It would be possible to mention some other pioneering expert systems here, such as Dipmeter Advisor, Dendral, and certain others, but the four examples given above serve to illustrate the main points about such systems. All of them held out tantalising prospects of opening up whole new areas of computer applications, and thus fired the imagination of academics and those interested in the commercial exploitation of such systems alike. They held out a good deal of promise, particularly XCON which was already a profitable proposition, and they sketched out the basic architecture and scope of expert systems. The quesion thus arises as to why the anticipated flood of successful commercial expert system applications did not arrive.

One feature in common to all the above examples is that they all attempted to tackle big problems, exercising highly skilled people. This was because such systems, needing to be developed from scratch and employing technology slightly beyond the leading edge, were extremely expensive undertakings. This meant in turn that to be worth doing they had to address applications with a high potential payoff, which meant those which currently used the skills of expensive people. Two problems arose from this: first, it meant that the tasks had a very high degree of difficulty so that the technology was in some ways trying to run before it could crawl. Second, there was an understandable degree of reluctance to put trust in the conclusions of such systems. A mistaken diagnosis from MYCIN could be a matter of life and death, and a wrong guess from PROSPECTOR could lead to a disastrously expensive exploration of an unfruitful site. Perhaps it is not coincidental that the successful system XCON differed from the others in the lesser consequences of mistakes from the system, which allowed its judgements to be followed without too many ill-effects if they were wrong.

The failure of expert systems on the model of these early systems to become widespread in real environments is therefore largely explicable because of the scale of investment required when compared with the probable benefits of the results. Expert systems development therefore took a rather different tack, centring on the development of *expert system shells*.

9.6. Expert system shells

The basic idea of an expert system shell follows from the separation of the

158 Knowledge Representation

knowledge base and the inference engine. Given that the two are distinct, it can be imagined that the knowledge base might be manipulated by a different inference engine, or that the inference engine could be used to manipulate a different knowledge base. In practice, the first of these possibilities is not easy to realise: the formalism used by the knowledge base is targeted at the intended inference engine, and the knowledge base is likely, for the reasons discussed above, to contain non-declarative features designed to exploit particular features of the inference engine, such as the conflict resolution strategy to be used. The second, however, does appear very possible, and gives the attractive promise of allowing a series of expert systems to be constructed sharing many common facilities and differing only by being supplied with different knowledge appropriate to the task they are to perform. Moreover, because the architecture of expert systems is such that the interface facilities communicate with the knowledge base only through the inference engine, leaving the inference engine unchanged means that we can immediately use the facilities for interrogating the user and providing explanations.

Essentially the idea then is that we can first remove the domain knowledge from an existing expert system, to leave us with an expert system shell, comprising the inference engine and the interface components of the original system. A new knowledge base can now be constructed for a new domain and added to this shell to give us a new expert system capable of working in a new domain.

This idea was first put into practice within the MYCIN project. Taking MYCIN as a starting point the knowledge pertaining to diseases of the blood and their treatment was removed, leaving a shell known as EMYCIN (variously believed to stand for "empty MYCIN" or "essential MYCIN"). This shell was then used in a variety of further projects to construct a number of expert systems, mostly in the medical domain, most notably PUFF, an expert system for the diagnosis of respiratory diseases. These experiments confirmed the hypothesis that the inference engine aspects of an expert system, and hence the knowledge representation formalism it manipulated, did not need to be developed afresh for every new domain, and so the shell approach seemed to offer a good prospect of simplifying expert system construction.

9.6.1. Impact of shells

The impact of the idea of expert system shells was dramatic. Whereas previously expert systems had been undertakings requiring an immense amount of investment because they had to be designed and built from scratch, using a shell could put them within the reach of a much wider market. Users could purchase a shell and use their own expertise to supply the domain knowledge and so get an expert system for their own purposes. This meant that companies which had been attracted by the prospective advantages of expert systems but which had been daunted by the costs of exploring the technology, and which perhaps could not identify any application with a sufficient high prospective payback, could now acquire a shell and use in-house expertise to explore applications at a relatively small cost. These applications could be much more modest in their ambition, and produce demonstrations of the applicability of the technology in a relatively short time.

This meant that there was a sudden growth in the application of expert systems technology to applications with a much smaller scale, and involving far less difficult questions of judgement than had been necessary with the earlier expert systems which had had to be built from scratch. Also, the cheapness of these demonstrations meant that far more of them could be produced, and the potential for expert systems across a far wider range of tasks could be explored.

In terms of cost-benefit, several of these applications proved to be potentially profitable; whereas before shells applications had needed to be large, now it was possible to produce a useful expert system which was on a small scale doing a constrained task. Such applications could not have hoped to justify the expense of developing an inference engine, but the modest knowledge base they required was feasible. Moreover, the tasks these systems addressed were, of course, easier to tackle than the large tasks required to justify an entire expert system development.

Thus the impact of shells was to open up a wide range of applications to the technology and to lower the demands placed on expert systems. As this became clear, expert systems shells became highly attractive products for those who wished to market them. Here the development costs could be spread over a high volume of sales, and provided the price could be kept reasonably low, there was a large market for these products. Moreover, what increasingly became seen as the major stumbling block to expert systems development, namely the construction of an effective knowledge base, was passed away from the shell producer on to its customer. For these reasons expert systems shells proliferated, and today there is an enormous range of products. In the early days shells exhibited a wide variety of approaches, based on one of the model expert systems, usually MYCIN or PROSPECTOR. Gradually, however, a greater uniformity of approach became evident, and we can now say that shells typically contain a fairly common set of features. These features will be outlined in the next section.

9.7. Typical facilities of an expert system shell

As has been indicated above, expert system shells offer a variety of facilities. However, the basic style of the majority of such systems has tended to follow the sort of model suggested by MYCIN. Essentially this means that they tend to be predominantly rule-based, largely because this was found to be a model readily understandable by the users at which such systems were aimed, and to take consultation with an expert as the model of interaction. What these decisions mean in terms of the facilities provided will be discussed below.

9.7.1. User interface

Expert systems presuppose a model of interaction in which a user consults the system. This requires that user and system be able to interact. Thus the expert system will incorporate a means of obtaining a goal from the user, and then use its knowledge base to pose questions to the user to elicit the information required to provide an answer to the user's initial question. Some expert system shells will attempt to construct the questions in terms of the predicate names used in the knowledge base (e.g. early versions of APES [9]). In practice this was found to produce either a very stilted, or in worst cases incomprehensible, dialogue, or a knowledge base with cumbersome predicate names, and most later systems require a string of text to be associated with a particular question. This enables the programmer to supply a piece of *canned text* (a term used to describe a piece of text provided as a unit by the system builder rather than being constructed when required by the system) which will be shown to the user of the system when the value for a particular attribute is required. This is a very simple mechanism, but greatly improves the dialogue from the user perspective without placing too great a burden on the system builder.

Also required is a means of selecting the questions to ask the user. Very often this will be determined by the operation of the inference engine, and a question will be posed only when the piece of information is required by the system. In some more sophisticated systems a scheme of prioritisation of questions might be made available.

Apart from managing the dialogue using the above facilities, the other important aspect of the interface is the provision of explanations. Great emphasis has been laid on the ability of the expert system to explain its conclusions, both because such explanations can aid the user in understanding the output of the system and because without explanation users tend to place little confidence in this output. There are essentially two explanation facilities which come with (almost) all expert systems shells of this type, both of which derive from MYCIN, and a third which goes a little beyond these but which has been found popular with users.

The two MYCIN-related explanation features are the abiliy for the user to interrupt the dialogue to ask why a particular question is being asked, and for the user to ask how a particular conclusion was reached when the expert system returns an answer to a question. Both of these, as has been explained in Section 9.4.3, are easily implemented within a rule-based expert system from an examination of the rules being fired.

The extension to the MYCIN-based explanation facilities, often termed a *Whatif?* explanation, allows the user to interrupt the dialogue to find out what the consequences are of giving a particular answer to the question being posed and so helps him to decide on the answer he wishes to give. The facility is interesting because the need for it did not emerge in MYCIN, but was seen as expert systems shells began to be applied to a wider variety of applications. In the situations for which MYCIN was designed, the user was unable to alter his answers according to the result he wished to achieve, because it was envisaged that he would have a sample and that this sample would have certain properties. In some of the newer applications, however, the user can have control over the facts of the situation about which he is being questioned. For example, suppose we have an expert system which is offering advice on certain welfare benefits; it may be necessary to know whether or not the user is retired. Now under UK social security law, a user may choose whether or not he wishes to be treated as retired, and indeed may be consulting the system to find out what he should decide on this matter. If he is asked whether he is retired or not then, in the absence of a Whatif? facility, all he can do is answer one way and, on reaching the end of the consultation, run the system again to find out if his situation would improve had he given a different answer. The Whatif? facility means that this laborious procedure is avoided. A second motivation for this facility was that sometimes the user could not understand the question sufficiently to know what response to give: here, if he is able to follow the consequences of the answer forward he may well get a better appreciation of the meaning of the question. This then was a useful extension to the armoury of expert systems explanation types, deriving from the experimentation permitted by the increasing availability of expert system shells.

9.7.2. Knowledge base

Of course, given the nature of an expert system shell, nothing in the way of a knowledge base is provided, since it is up to the purchaser of the shell to provide the knowledge that completes the expert system. Equally, given that the inference engine is supplied, the knowledge base supplied by the system builder needs to be of a form that the inference engine can manipulate. Thus an expert system will supply not a knowledge base but a formalism in which the knowledge making up the knowledge base will be encoded.

This usually addresses two issues: the representation of facts, and the representation of rules (the overwhelming majority of shells being rule-based). Facts are typically based on the entity–attribute–value triples discussed in Section 5.1.1, although most shells simplify this to attribute–value pairs. This is very helpful, since it means that the underlying logic of the system need be no more than propositional, greatly simplifying the design and efficiency of the inference engine. This is, however, a fairly stringent restriction, greatly decreasing the expressive power of such a formalism, since it is impossible to describe relationships between more than one entity in a way which is even-handed between the entities. In practice, however, most expert systems find themselves able to work within this restriction, since, in the applications to which such shells are put, there is typically only a single entity under consideration in any given consultation. Thus a medical dignosis system is looking at a single patient, a plant-classification system at a single specimen, and a benefits-advice system at a single potential claimant. In each case therefore it can be plausibly maintained that had there been the facility to refer to an entity in the facts, the same entity would have been referred to throughout, and thus the attribute–value pairs may be construed as entity–attribute–value triples, with the entity implicit. This simplification does indeed work well for most of the time across a large number of applications, although it does on occasion require working around. To take an example: benefits advice often requires information as to the circumstances of the spouse of the claimant, and this would naturally be expressed using two entity references, but it can be expressed by having something like 'ageOfSpouse' as an attribute of the claimant. For many of the simple applications to which these shells were put, this inelegance and inflexibility was tolerable, and was well compensated by the gains resulting from the simplification of the logic. Some systems do, however, allow the use of multiple entity references.

Given a formalism for facts, rules could be constructed out of these facts together with some simple logical connectives. The typical format is IF some conjunction of facts THEN some conjunction of conclusion, which is to say the standard production-rule form. Shells will normally place some restriction on the number of facts allowed in the conditions and conclusions parts of the rule.

These are the absolute basic facilities. In additions, shells will offer ways of breaking the knowledge base up into modules (to aid in reading the knowledge base, to increase the efficiency of execution and, especially where the system is to be run on a microcomputer, to allow the knowledge base to fit into the memory), and perhaps ways of grouping facts together into something resembling frames. Normally the knowledge base will be designed to be constructed using a standard text editor, but some shells may provide special editors.

9.7.3. Inference engine

Perhaps the most important part of the expert system shell is the inference engine. This will usually be in the form of a rule interpreter. By far and away the most common form of interpreter offered is one which backward chains from a goal, asking the user to supply any facts neither in nor provable from its database. This form of inference engine has become standard largely because it fits in naturally with the needs of an expert system, particularly one built along the lines of a consultation. It has the advantage of only traversing paths on the way to a potential solution, and providing a fairly natural way of generating questions. Some shells do give users some option over the control of the inference; in particular, some systems work well if some initial facts are solicited from the user and forward chaining is performed from them before backward chaining towards the goal. Nevertheless, backward chaining remains the most prevalent style for expert system shells, and gives rise to the typical interaction with such shells.

Most shells also offer something by way of treating uncertainty in rules and facts. This may be as simple as allowing a response of "don't know" to a question (whereupon the information will simply not be used and so any rule containing this information or its negation will not fire), but more usually will involve the propagation of certainty factors through rules. Again MYCIN's treatment of uncertainty provides the standard model for how this is to be done; a conjunction of conditions is given a certainty equal to the minimum certainty factor of its component conditions, and the certainty of the conclusion is the product of this certainty factor and the certainty factor of the rule. Although uncertainty handling is offered by most shells, and was originally seen as a crucial component of expert systems, successful expert systems have, in fact, tended to avoid use of uncertainty. The treatment of uncertainty will be discussed more fully in Section 10.4.

9.7.4. Escape to underlying system

A final component of expert system shells that should not be overlooked is the ability to escape to the underlying system. The rigid formalism imposed by the shell was often found to be unduly inhibiting by the system builders, and there was also a desire to enhance the interface to the system by, for example, the provision of graphics. Thus it was found necessary to give the system builders access to the underlying system so that certain routines could be encoded in a conventional language and invoked from the expert system shell. Of course, any use of this feature will inevitably mean a reduction in the declarativeness of the knowledge base.

9.8. Trends in expert systems

The significance of the impact of shells in changing the appreciation of what tasks are suitable for expert systems technology and the view of what expert systems are cannot be overestimated. The original idea underlying expert systems such as MYCIN (and, but more doubtfully, XCON), was firmly within the field of artificial intelligence, and the builders of these early systems were interested in capturing features of the behaviour of highly skilled people working near the limits of their capabilities. Primarily this was motivated by genuine research aims. The idea, however, was commercially attractive, and the real application of such systems was envisaged. But the commercial application of systems with this kind of ambition represented an over-optimistic view of the capabilities of such systems within the present state of the art. Successful performance was needed, however, if such systems were to justify their expense, and it appeared that expert systems would move back into the laboratory.

The experience of XCON, however, held out a rather different prospect. That system did capture the performance of highly skilled people, and was a paying proposition. With hindsight we can see that the significant feature, however, was that the task addressed was not one which stretched these experts, but rather one that they found routine and resented because it kept them from tasks which did fully utilise their expertise and which therefore they regarded as preferable. Here the payoff came from the fact that skilled resource was needed, and that the task had a high volume.

The widespread availability of shells meant that this kind of task could be experimented with on a large scale across a broad range of organisations, and this meant that more of these routine tasks could be incorporated into expert systems. Although the payoff was much smaller, these systems could be shown to work, and to make a worthwhile contribution, particularly in freeing experts for more congenial and demanding tasks where their judgement and expertise is essential to the task.

As the technology became more widespread, understanding of both the scope and limits of potential expert systems applications, and how to go about building them, grew and there are now many paying expert systems in operation. All this is, however, possible only because the cost of expert systems development has fallen dramatically through the use of shells. Such systems are very different from the original conception of expert systems, and it is probably true to say that they lie outside the field of AI, and belong rather to mainstream computing, so that rule-based programming using an expert system shell is an alternative to using a conventional or fourth-generation language, rather than an entirely separate technology. In many ways, all these points were latent in the success of XCON, which supported a task of this sort, but the message of XCON's success was somewhat blurred by the rather more exciting claims of expert systems emerging from the research community.

9.8.1. What kind of tasks can we use expert systems for?

This change in appreciation of the nature of expert systems and their practical applications can lead us to the following suggestions as to what characteristics a task amenable to being supported by an expert system is likely to have. These differ in many ways from the features of tasks suitable for this treatment that were identified in the early 1980s when expert systems were first coming on the scene, and are worth noting.

First, the knowledge needs to be readily available, either in the form of an expert prepared to supply it, or in some written form such as the instructions and guidance supplied to clerical workers who might be currently performing the task. This contrasts with the early view that the expert system task would depend on a good deal of tacit knowledge, possessed but not articulated, by experts, which would need to be painstakingly elicited and articulated by a knowledge engineer. This latter view led to the so-called knowledge elicitation bottleneck, which was seen as a major hurdle to expert systems development. The point is that uncovering the tacit knowledge required to perform the hard tasks carried out by an expert is a difficult if not impossible task, but that workable expert systems can be built for lesser tasks where this degree of teasing out of the required knowledge is unnecessary. This being so, we can say that if the expert is unaware of the knowledge that is guiding him, the task is probably too complicated.

Following on from that we can say that it is desirable that the expert systems task should be well understood, at least by those who do it. The early expert systems task was seen as one where the data considered was uncertain, the relevance of particular items of data debatable, and the conclusions reached such that experts might disagree as to their validity. Again, these can now be viewed as characterising a task beyond the routine and too difficult for current practical expert systems. The better task is one for which someone (the expert) is confident as to what data should be considered, and what effect this data has on the conclusion he will reach, and for which he would expect almost universal concurrence in his decisions from other experts. This is what characterises the routine task.

The knowledge required to solve the task should, however, be fairly

166   Knowledge Representation

extensive, since otherwise the task becomes trivial. If the knowledge to be encoded in the expert system could be taught easily to some moderately able person, then an expert system will not be justified. How extensive the knowledge must be is a matter of judgement (probably beyond the capacity of an expert system), and depends on the nature of the domain and factors such as the extent to which the various parts of the knowledge base will interact, and the existence of some cases which fall into a category which requires some different treatment but which is encountered only rarely. This last is particularly important, since it is often forgotten that whilst applying a rule may be trivial, remembering its existence and determining its applicability may not be.

9.8.2. What are the advantages of expert systems?

What advantage, then, could be expected from adopting an expert systems solution to such a task? First, since it will be obvious that, if a number of people within an organisation are performing a task, some will be better at it than others, we can see the incorporation of the knowledge required for the task into the knowledge base of an expert system as a means of making the best knowledge available to all those performing the task. Less obviously, but perhaps more importantly, even the best of experts will have off days when they will perform the task less well than they are capable of doing. The knowledge in the expert system will represent their considered judgement as to what they should they do, and using the expert system will mean that they are using their best knowledge and so help to avoid oversights and errors. Related to this, the use of expert systems will help to ensure that the decisions taken are consistent, both amongst different people, and for a given person at different times. Provided the decisions are good, this will will be an advantage.

Second, the expert system will help to ensure that the knowledge is applied accurately. Again, the person performing the task may be well aware of a particular rule, but on occasion fail to apply it correctly. This can be seen clearly when the task involves some element of arithmetic calculation. Of course, a real expert is perfectly capable of adding and multiplying, but equally he will be bound to make the occasional error. But this does not just apply to arithmetic, which could be handled as well in a conventional system, but to the application of symbolic rules as well.

The factors point to the routine task being performed better and more accurately, but we would also expect that the introduction of an expert system would enable the tasks to be performed more quickly, or by a less able person. This means that the expert can concentrate on more appropriate work, giving more time to the really testing tasks, and so performing those better as well. As a side-effect, this ought to make the job more pleasant for him as well. Also, if the expert system can provide support which enables less able people to do a task previously beyond their capacity, they may find that this enhances job satisfaction as well. (All this is the up-side of expert systems: concerns about deskilling and loss of individual consideration of each case are real and worth thinking about too.)

Lastly, some of the old claims of expert systems, that they allow for prototyping, and provide a faster route to building the application, that they are capable of being maintained more easily, that they can be gradually refined and extended, can still be made, although it is perhaps too early to say whether they have been really proved in practice.

9.9. Examples of current expert systems

I shall end this chapter by briefly looking at two examples of these new style expert systems built using shells and with no pretensions to capture great expertise.

9.9.1. Latent damage advisor

As an example of the kind of system that would result from this approach, we can first consider the Latent Damage Advisor, built using the CRYSTAL shell, by Richard Susskind and Philip Capper [10]. This system is designed to support a qualified lawyer who is not an expert specifically on the law of Latent Damage to come to a view of a case in this area. The law on Latent Damage is complex and rarely encountered, except by a few specialists. It is, however, closely related to the law on negligence, with which most qualified solicitors will be familiar. The system thus acts as a guide through the law pertinent to Latent Damage, to be used by a competent lawyer familiar with the law of negligence. The target user group means that the system need not deal with the really hard aspects of the cases: where real expertise and judgement with respect to law are needed the onus is placed upon the user. The user can see what questions he must answer, and what the implications of his answers will be, but must do all the specifically legal reasoning himself. None the less the advantages of the system are still considerable. The claim is made in Ref. 11 that

> It is estimated that a competent lawyer would take five to ten hours to understand the Act and its implications. ... By using the Latent Damage System, however, the lawyer can find solutions to latent damage problems in about five to ten minutes. The system is an "intelligent assistant"—it guides a user through all and only those legal rules which bear on the problem at hand.

If anything, this understates the usefulness of the system: the expert in question, Philip Capper, who is an acknowledged expert in the field, and an author of the first book on the statute, found himself surprised on occasions by the output of the system, and that the system was, on reflection, right. This suggests that five to ten hours' work would not result in an understanding as good as that provided by using the LDA.

9.9.2. Pensions advice

A second example is provided by the Retirement Pensions Forecast Advisor (RPFA) developed by Arthur Andersen Management Consultants for the UK Department of Social Security [12]. This system again used a commercially available shell, this time, the Aion Development System. This system was designed to support clerks advising people (by letter) as to whether they could expect to receive a retirement pension when they attained the appropriate age, and if so, at what rate they could expect to be paid. This advice requires calculations based on the employment history and contributions record of the individual concerned, and other factors such as their marital history, children and absences abroad. Naturally this differs from individual to individual. Further complications exist since certain options are available to some people: thus a married woman may substitute her husband's contribution record for her own in certain circumstances. The manual system suffered from a number of defects, most notably the accuracy of the advice, the time taken to produce it, the cost of producing it, and the quality of the letter sent out. These defects arose not because some cases were hard, in the sense that ground-breaking decisions needed to be made, but because of the complexity of the rules, the multiplicity of their interactions, and the fact that some rarely applicable rules tended to be overlooked. There were, however, no problems in determining whether particular rules were applicable in a given case, nor in determining the applicable facts. It was therefore possible to devise a set of rules which could handle all the routine cases with confidence. There remained, however, a number of cases which were non-routine, because the applicant had worked for a period abroad. No attempt was made to automate these cases; and the presence of a straightforward criterion to separate the easy cases from those requiring detailed examination meant that there was little degradation in the utility of the system. The system is determinate and requires no judgement; it could have been built using a fourth-generation language, or in a conventional imperative language. The expert systems route was chosen, however, because it provided a speedier implementation (important since savings are said to be well over £1 million per annum), and because it was felt to be more maintainable. The RPFA system therefore provides an excellent example of a system for routine application of the law to decide cases where there are a large number of routine cases, and where these can easily be identified. Note that it is the large volume of cases that need to be decided (currently 300 000 + a year) that makes this system viable.

These systems represent expert systems successes, and typify the more modest ambitions of current expert systems. The use of Advisor in both their titles is perhaps significant; the desire to supplant or equal the performance of experts has gone, to be replaced by the notion of providing support for the routine tasks.

References

1. Samuel, A.L., "Some studies in machine learning using the game of checkers", in *Computers and Thought* (ed. E.A. Fiegenbaum and J. Feldman), McGraw-Hill, New York, 1963.
2. Buchanan, B.G. and Shortliffe, E.H., *Rule Based Expert Systems*, Addison-Wesley, Reading, Mass., 1984, p. xii.
3. Miller, P.L., *A Critiquing Approach to Expert Computer Advice: Attending*, Pitman Advanced Publishing Program, Boston, 1984.
4. Buchanan and Shortliffe, *op. cit.*
5. McDermott, J., "R1: a rule based configurer of computer systems", *Artificial Intelligence*, **19**, 39–88 (1982).
6. Alty, J.L. and Coombes, M.J., *Expert Systems: Concepts and Examples*, NCC Publications, Manchester, 1984.
7. Pople, H.E. Jr., Myers, J.D. and Miller, R.A., "DIALOG: A model of diagnostic logic for internal medicine", *Proceedings of the 4th International Joint Conference on AI*. [DIALOG was an early name for INTERNIST].
8. Jackson, P., *Introduction to Expert Systems*, Addison-Wesley, Reading, Mass., 1986.
9. Hammond P. and Sergot M.J., "A PROLOG shell for logic based expert systems", *Proceedings of Expert Systems 83*, pp. 95–104. [APES is marketed by Logic Based Systems].
10. Capper, P.N. and Susskind, R.E., *Latent Damage Law—The Expert System*, Butterworths, London, 1988.
11. Susskind, R.E., "Government applications of expert systems in law", in *KBS in Government 88* (ed. P. Duffin), Blenheim Online, Pinner, 1988, pp. 185–204.
12. Spirgel-Sinclair, S., "The DHSS retirement pension forecast and advice system", in *KBS in Government 88* (ed. P. Duffin), Blenheim Online, Pinner, 1988, pp. 89–106.

10
Some Issues in Knowledge Representation

Now that we have looked at the standard knowledge representation paradigms separately, it is time to draw the strands together and make some comparisons between them. Arguments of this sort sometimes become unprofitable because people become very attached to their favoured representation, and so given to dogmatic assertion, but there are a number of things which are worth saying. In this chapter I shall consider some points of similarity between the various paradigms, namely production rules, structured representations as exemplified by frame systems incorporating an inheritance mechanism, and first-order predicate calculus, and then go on to discuss a number of areas where the expressive power of these representations seems to be inadequate, and extensions that have been proposed to address the resultant problems.

10.1 Similarities between the paradigms

The first thing to note is similarity. Although the representations have very different surface forms, the entity–attribute–value triples found in production systems, the instance–slot–filler notation of the frame system, and relations with two parameters found in predicate logic, all express precisely the same information, namely that a binary relation holds between two objects in the domain. Predications are, as has always been recognised by formal logic, the typical form of an assertion of a fact about the world, and much knowledge consists of such predications. Thus it is unsurprising that there should be this underlying commonality. Taken as means of describing facts by making predications, therefore, the three paradigms are essentially equivalent in expressive power. There are stylistic differences in that the production and frame systems lead us to a more object-centred view, whereas predicate logic is more relation-centred, but there is a simple mapping between these two perspectives. At this level, therefore, there is little to discriminate the paradigms, except personal taste in that a person may find one of them more congenial than the others.

One extra thing that predicate logic does offer is that within this paradigm, unlike the other two, we are not confined to expressing binary relations. Thus if we want to represent the three roles in a giving event, we can use a three-place relation for giving, whereas in the other representations we must introduce an extra entity to represent the giving event and produce an equivalent set of binary relations. The set of binary relations produced is often more difficult to comprehend at a glance than the sensibly grouped information given by a relation with more places. Frames are in part a response to this difficulty: they are useful because they do allow related information to be grouped together; thus the slots in the frame representing a giving event are rather like the roles in the many-placed relation likely to be used in predicate logic. This grouping together of information is often cited as an exclusive advantage of frame systems; one advocate of the representation writes [1]

> 'So I would claim the "bundling" aspect of frames,
> even for those cases where there is a mapping onto
> predicate calculus . . . has been one of the major
> practical attractions of the frame idea.

This is correct when frames are compared with the entity–attribute–value triples used in production systems, but ignores the fact that the ability to use many-placed relations also provides a "bundling" aspect for predicate logic. However, when a large number of things need to be bundled together the point is probably correct: a frame with twenty slots is more immediately understandable than a twenty-place relation.

This similarity does not, however, extend to the expression of knowledge about the relationships between these relations. In predicate logic this knowledge is expressed by clauses where the connectives used have a fixed truth functional meaning. It might appear that production rules are similar in that the IF . . . THEN used there seems to correspond to the material implications of predicate logic. It is, however, important to recognise that this is not necessarily so. For it is not true that the relation between the conditions and conclusions of a production rule is entailment. For the conditions may be satisfied and the conclusions not be capable of being drawn; for them to be drawn the rule must fire, and whether or not a given rule can fire will depend on the other rules in production memory and the conflict resolution strategy employed by the rule interpreter. Further, the nature of working memory is such that conclusions may be deleted from it, so that a rule may be applicable in one cycle and not in the next. We cannot therefore identify the triples in working memory with the timeless propositions of logic. The precise meaning of the connection between conditions and actions in production rules is therefore unclear, whereas in predicate logic the meaning of material implication is well defined. This may be an advantage: the lack of precision enables a degree of flexibility to be incorporated, and rules expressing something which would be untrue were the rule read as a material implication may be written without harmful effects, provided the behaviour of the program when the knowledge is executed is correct. The situation with regard to structured objects is even more extreme: there the relations between relations are expressed only in the way in which the various link types are manipulated. This again offers a great deal of flexibility, but can lead to there being no possibility of a uniform declarative reading of what is expressed.

In the remainder of this chapter I will discuss a number of issues relating to knowledge representation as understood in the context of the various paradigms we have considered here, mostly deriving from a desire to extend the expressive power of the notations to cope with knowledge that goes beyond simple predication of relationships. Most of these topics will open up research issues, as we will be coming up against the limits of these paradigms. None the less, it is well worth while to be aware of these issues, as they are both indicative as to the limits of state of the art of well-understood methods of knowledge representation, as represented by these paradigms, and indicative of the problems that are likely to be faced when encoding knowledge relating to a particular domain.

10.2. Expressiveness of Horn Clauses

The first of these issues is whether the Horn Clause subset of first-order predicate calculus is sufficiently expressive to provide us with a satisfactory means of knowledge representation. This is an important issue because the exploitation of logic as a knowledge representation paradigm has largely been in the context of Horn Clauses, owing to the availability of PROLOG. Certainly the restriction to the Horn Clause subset makes the representation computationally much more tractable, but this will be of no avail if the restriction means that the representation ceases to be expressively adequate.

The restrictions placed on us by the need to use only Horn Clauses are essentially two; that we can only have one literal on the left-hand side of the clause, and that we cannot have any negated literals. That these restrictions taken at face value would be too inexpressive for practical purposes can be seen by considering an example as simple as expressing the fact that all human beings are either male or female. In first-order predicate calculus we would write

$$\text{FOP1 } (\forall x)(\text{human}(x) \rightarrow \text{male}(x) \lor \text{female}(x))$$

In clausal form this would become

$$\text{CF1 male}(x) \lor \text{female}(x) \leftarrow \text{human}(x)$$

putting a disjunction on the left-hand side. The restriction, therefore, is clearly intolerable, since we cannot live with the inability to express this kind of knowledge. A fairly standard way of reasoning is to deduce that several things are possible, and then to prove that all but one of these candidate solutions are inconsistent with the facts (which is to say that we generate a set of hypotheses and exclude all but one of them): this would be entirely impossible using the Horn Clause subset. Computational tractability will not allow us these disjunctive conclusions, so what can we do?

10.2.1. Negation in logic programming

If we look at CF1 we can see that we could get the desired inference by moving one of the literals on the LHS to the RHS and changing its sign. This is in conformity with the way of reasoning mentioned above; firstly establishing some conditions which entail the disjunction and then falsifying all but one of the disjuncts. So we could represent FOP1 (assuming that we were allowed negations on the RHS) as

HC1 male(x) ← human(x) and not female(x).

This suggests that what we need is a way of deciding negations in the bodies of Horn Clauses. We cannot use classical negation here, because that would involve having a means of demonstrating the falsity of a literal, which would in turn demand the ability to write negated literals on the LHS of Horn Clauses, which we know we cannot do. We therefore need to find some other way to demonstrate the falsity of a clause. What we need then is a notion of negation which, whilst not being classical negation, is sufficiently close to classical negation to allow the transformation between CF1 and HC1. The notion adopted within logic programming (and embodied in PROLOG) is known as *negation as failure*. The basic idea here is that a literal is taken to be false just in case that it cannot be shown to be true. Thus to establish the falsity of a literal we attempt to show that it is true, and if we fail to do so, we take it as shown that the literal is false. This has in common with classical negation that it ensures that every literal has one and only one truth value, true or false, but it is significantly different in that whilst, classically, it is possible that neither a proposition nor its negation is provable, here the unprovability of a proposition is taken as sufficient (and necessary) to establish its falsehood. Put briefly this means that classically we can say that a proposition is either true or false, but we neither know nor can prove which; but if we interpret negation as failure this state of ignorance is impossible, since anything not known or provable will be taken as false.

10.2.2. The closed world assumption

What this treatment of negation is doing, therefore, is to assume that the truth value of every literal is known: either we know it to be true, or we can show it to be true, or else it is false. Making this assumption, known as the *closed world assumption*, enables negation as failure to be taken as equivalent to classical negation, as has been shown by Clark [2]. The question therefore arises as to how realistic this assumption is. We may think it a reasonable constraint in the case of facts, but we should note that it also has an implication with regard to rules. Thus take a simple clause of the form

>   MD1 has__runny__nose(X):-has__cold(X).

On the face of it, this expresses a truth, because having a cold is indeed a sufficient condition for having a runny nose. But there are, of course, many other causes of a runny nose—hay fever, for example. We can start to list these other causes by adding clauses such as

>   MD2 has__runny__nose(X):-has__hay__fever(X).

However, no matter how many such clauses we add, we will never be doing more than expressing sufficient conditions. But if we make the closed world assumption, we are in effect saying that the sufficient conditions that we have enumerated provide, *when taken together*, the necessary conditions for the truth of the predicate. If our program consists of no more than MD1 and MD2, we will be entitled to conclude from the joint failure of has__cold(john) and has__hay__fever(john) that has__runny__nose(john) is false only if we make the assumption that there are no other explanations for John's runny rose. The point is that if we make the closed world assumption we not only commit ourselves to saying that we will be told all the relevant facts, but that when we write clauses allowing things to be deduced, we will write an exhaustive set of clauses, and that no other ways of deducing these things exist. This idea of taking the set of clauses for a predicate as expressing necessary as well as sufficient conditions for the truth of that predicate is essential to Clark's demonstration of the soundness in classical terms of negation as failure, and is known as the *completion* of the database. An important consequence is that we then have no way of writing a set of conditions that are sufficient only: negation as failure means that we cannot make a distinction between providing a set of sufficient conditions and a set of jointly necessary and severally sufficient conditions.

The point may be made formally. When we write a set of clauses

> $P \leftarrow Q \& R$
> $P \leftarrow S \& T$
> $P \leftarrow U \& V$

we do not rule out the possible model where P is true and none of Q & R, S & T and U & V are true. If we take negation as failure, however, we are effectively adding a further clause

$$P \rightarrow (Q \& R) \vee (S \& T) \vee (U \& V)$$

so that the clauses taken together are equivalent to

$$P \leftrightarrow (Q \& R) \vee (S \& T) \vee (U \& V)$$

In some domains this will be a realistic assumption. When our clauses represent definitions, such as a person is bachelor if he is male and unmarried, the assumption is reasonable. So too in the realm of mathematics, where to say that a plane figure is a triangle if it has three straight sides, again expresses both necessary and sufficient conditions. But where we are dealing with more empirical domains, such as the cause of a runny rose, or the failure to fly of a bird, the assumption looks a good deal less plausible. Consider

```
flies(X):-bird(X),not flightless(X).
flightless(X):-kiwi(X).
flightless(X):-penguin(X).
flightless(X):-ostrich(X).
```

This program could be made to be satisfactory so long as we are considering the disposition to fly, since we could, in theory, list all the species of bird which are flightless. If, however, we were rather interested in whether a particular bird was capable of flight at a particular moment (perhaps because we wanted to know whether we should put it in a cage with a roof on it), we would need to have some extra clauses such as

```
flightless(X):-has__broken__wing(X).
flightless(X):-feet__set__in__concrete(X).
```

Here we would need to anticipate every circumstance which could render a bird flightless normally capable of flight, and we could never be sure that we had done this exhaustively. (The second clause derives from an example of Minsky's which purported to show a circumstances which no one would ever consider when writing an AI program. I shall not attempt to give such an example since to do so is to make it one everyone will think of (as I did with Minsky's). Let us just say that it is obvious in such a case that it is impossible to envisage every eventuality.)

To summarise the argument so far, if we are to have sufficient expressiveness in Horn Clauses, we must, at a minimum, extend them so as to allow for negated conditions in their bodies. To do so in a computationally tractable way, we must interpret negation as failure, committing ourselves to the closed world assumption so that we can rely on the completion of the database to give the required correspondence to classical negation. This seems plausible for some domains, notably those where the knowledge derives from definitions, and implausible for others, where the knowledge is largely in the form of empirical observation of sufficient conditions.

10.2.3. Limitations of negation as failure

Even if we are in a domain which is suitable for taking negation as failure, there is a further problem. Suppose we wish to represent CF1. This cannot be done fully by HC1, since whilst that will enable us to derive that X is male from X is human and the falsity of X is female, it will not, as CF1 would, allow us to derive that X is female from X is human and the falsity of X is male. We would therefore need two clauses, HC1 and

HC2 female(X):- human(X) and not male(X).

But we cannot expect HC1 and HC2 to work together satisfactorily as a logic program. For suppose we wished to discover the sex of Hilary, without any other information; then we would try to solve HC1, and discover that we needed to demonstrate the falsity of male(X). This can only be shown if we fail to show that X is female from HC2. This in turn requires a further use of HC1, which again calls HC2 and a fatal loop is established.

This means that when we wish to represent something that would naturally use a disjunctive conclusion, and instead have to write it with all but one of the disjuncts negated in the body of the clause, we have to make a choice as to which of the original disjuncts we will leave as the head of the clause. This in turn means that when we write our representation we must know in advance which deductions we will wish to make, that is, the use to which we are going to put the representation. This is a significant deviation from the declarative ideal, in which we aim to represent the required knowledge neutrally, without regard to its use. In the case of CF1 above, we will have to choose which of HC1 and HC2 we are going to use and this choice will depend on whether we wish to be able to prove facts about a person being male, and rely on being told if the person is female, or to prove facts about a person being female and be told if that person is male. Representing a substantial amount of knowledge will require numerous small decisions of this sort, the cumulative effect of which will greatly effect the behaviour and capabilities of the system. In many cases the choice will be obvious; in a typical expert system it is fairly clear what questions we wish the system to answer, and what information it is reasonable to assume that the user will be in a position to provide, but even here it makes the task of transferring the knowledge to a different application less than straightforward.

So what are the conclusions for the expressiveness of Horn Clauses?

First that it is essential to have some means of treating negation so that negated literals can appear in the bodies of clauses, since otherwise the restriction to a single literal in the head of the clause would be too limiting. Provided with this facility we can represent most expressions that we want, but when we do so we cannot capture all the possible inferences in the original expression, but must make a choice as to which ones we want, in the light of the use to which the representation is to be put. Thus the expressive power of the representation is likely to be adequate for a given task provided we are aware of the restrictions that it imposes.

10.2.4. Other treatments of negation

In the discussion above the idea of negation as failure was introduced. The situation can be shown diagrammatically as in Fig. 10.1. Many people have found this situation unsatisfactory: where they are dealing with empirical domains they may well wish to have their rules for particular predicates treated as providing sufficient conditions only rather than necessary and sufficient conditions. One method of achieving this is to introduce a second predicate representing the absence of a relation. Suppose, for example we are writing a program which involves deciding whether or not someone is rich. We might have a set of clauses for rich such as

> rich(X):-drives(X,rolls-royce).
> rich(X):lives-in(X,hampstead).
> rich(X):-wears(X,gucci-shoes).

We would not wish, however, to treat anyone who lacked one of these signs of affluence as being not rich, as we would be forced to do if we took negation as failure. We could therefore, instead of using not rich(X) in the bodies of our clauses, use some other predicate, say notRich(X), and write some clauses giving conditions where it is applicable:

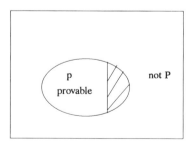

Figure 10.1 P is true within the whole ellipse, but, since P is not provable in the shaded region, negation by failure will return false even though it is true.

notRich(X):-drives(X,reliant).
notRich(X):-lives-in(X,tranmere).
notRich(X):-wears(X,plastic-shoes).

This can be useful, and does enable us to attempt a positive definition of the sufficient conditions under which a predicate is false. We should, however, be fully aware that we have not really defined a negation of rich(X). There is, in fact, no logical connection at all between the two predicates: the link is only an apparent one from the names we have used. What we really want to say, of course, is

– rich(X):-notRich(X)

but we cannot make this connection because of the inability to write negations on the left-hand side of the clause. Diagrammatically the situation is now as in Fig. 10.2. From this it is clear that there is a gap in our knowledge, and we will need to be aware of this so that we do not make incorrect conclusions about cases which fall into this gap.

An alternative to defining an additional predicate is to use an additional parameter to convey truth value. Thus we could write the above clauses as

rich(X,true):-drives(X,rolls-royce).
rich(X,true):-lives-in(X,hampstead).
rich(X,true):-wears(X,gucci-shoes).
rich(X,false):-drives(X,reliant).
rich(X,false):-lives-in(X,tranmere).
rich(X,false):-wears(X,plastic-shoes).

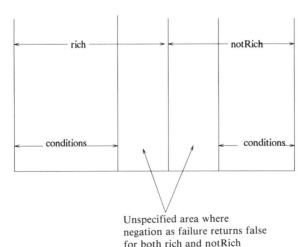

Unspecified area where negation as failure returns false for both rich and notRich

Figure 10.2

180   Knowledge Representation

This is in practice little different from the approach using two predicates, although it does have the advantage of bringing together the conditions for the truth and falsity of a predicate in a single procedure. Again, however, this is only truly satisfactory if we can cover every eventuality, and if we ensure that it is not possible for one of the conditions for rich(X,true) to be satisfied at the same time as one of the conditions for rich(X,false), since if this were so we would have a contradiction.

One possibility the extra parameter does open up, however, is using a third value, say "unknown" to apply to cases where none of the conditions above applies. Thus if we add a clause (using not as negation as failure),

rich(X,unknown):-not rich(X,true),not rich(X,false)

we will ensure that we get a value for the second parameter of rich in all cases, and have those cases where we do not know the financial status of a person drawn to our attention so that we can take some appropriate action.

To conclude, these alternative treatments of negation can have some use in particular circumstances where negation as failure is inappropriate, but they are not general and have to be specifically produced for each predicate we wish to treat in this way, whereas negation as failure can be implemented so as to apply to all predicates impartially. Nor are they equivalent to classical negation. Thus we are forced to consider negation carefully in each case, and use the treatment appropriate to our needs.

10.2.5. Negation in other paradigms

The problems of logic programming with negation have been discussed above. Such problems are most noticeable in this paradigm, because negation is a key feature of logic, and so its computational treatment comes to the fore. Similar problems do, however, exist with regard to the other paradigms. Take first semantic nets: there we have a very intuitive representation of the existence of a relationship, namely a link in the network. But what is the corresponding representation of the absence of a relationship? For if we simply omit the link, there will be no way of differentiating between the absence of a relationship and the absence of knowledge of the existence of the relationship. Thus if we take the absence of a link as meaning that the relationship does not exist, we are, in effect, making the closed world assumption and interpreting negation as failure. It could be proposed that we should have a special type of link for declaring that relationship does not hold, but that would be far too profligate of links. For if we have a "father-of" link, we would also need to have a "not-father-of" link which would need, when our information was complete, to connect a given person node to every other person node excepting the one who was the father of the person in question. This being so it does seem better to live with the implications of representing the absence of a relationship by the absence of a link. We can, however, provide some links with special inference procedures so that, for example, "adam father-of david" will return false only if some father-of link emerges from David to a node other than Adam, and unknown otherwise. This is particularly useful when we are dealing with one-to-one relations, but makes rather less sense with non-exclusive relations such as "child-of", and additionally complicates the inference mechanism that we are using to derive implications from our representation. One of the attractions of semantic nets is that the representation has a readily understandable diagrammatic form, and this tends to hide from us the fact that the inferences licensed by the net are not represented in this way. Hence, if we see that a father-of link runs between two individuals we would expect that the representation "knows" that it does not run between the child and another person: but this inference needs to be rendered operational. Thus there may well be more (or less) in a semantic net than meets the eye in studying the diagrammatic expression of the net.

Similar considerations apply to other structured object representations. The frame representation needs to be able to distinguish between the absence of a particular value from a slot and the knowledge that the value in question should be absent. In cases where the cardinality of a slot (the number of values which should occupy it) is known, a similar technique to that suggested above for one-to-one relations in semantic nets could be used. But many examples of indefinite cardinality abound: children-of, jobs-of, pets-of, to name but three. Thus the problem is difficult to solve in its generality.

If we turn to production rules, it might appear that the case is simpler. The negation as failure approach could easily be implemented by taking the negation of a triple to be true if a triple is absent from the facts in working memory. But there is nothing to stop us from distinguishing the facts in working memory between those known to be true and those known to be false, and, having done this, from writing rules to write to the part of working memory holding things known false, and so implementing the ability to draw negated conclusions, since there is no objection to negated triples in the conclusion part of production rules. Two points should be noted here. First, this could lead to a dangerous proliferation of the facts known to be false: when we learn that Jesse is the father of David, do we really want to add every other person father-of David to the things we know false? Second, there is a great danger of failing to enforce the logical connection between a fact and its negation: in principle every possible triple should be either something known true or something known false, and only one of these alternatives. This could well pose difficult problems for the management of working memory, or else collapse to something like the two-predicate approach suggested above for the logic paradigm.

To summarise, therefore, the explicit difficulties that arise with the treatment of negation in the logic paradigm should not be considered as an argument against the use of that paradigm. Very similar difficulties arise with the other paradigms as well, and the solution used in logic—to make the closed world assumption and interpret negation as failure—is the typical solution adopted, albeit often in a more *ad hoc* or implicit fashion in those other paradigms as well.

10.3. Non-monotonic reasoning

The second major issue that we need to consider is that of *non-monotonic reasoning*. This is an issue of the utmost importance since it denies one of the fundamental tenets of the classical logic approach, namely that deduction is an adequate model of human reasoning, and therefore a satisfactory basis for the manipulation of representations. Proponents of the need for some form of plausible reasoning argue that in practice many people draw conclusions which cannot be deductively supported, and which they are prepared to withdraw in the light of further information. This contravenes the principle of monotonicity, which is fundamental to classical logic, that if a certain conclusion can be drawn from a body of evidence, adding to that evidence cannot prevent that conclusion being drawn. That it is a feature of classical logic can be seen from the formal expression of the principle. A system is monotonic if given $KB \vdash P$ then $KB + \Delta \vdash P$, for any Δ. This must be true of classical logic since if we have a proof of P from KB, we can use that same proof as a proof of P from $KB + \Delta$. (Note that if Δ contains the negation of some proposition in KB this does not alter the provability of P, since within classical logic anything whatsoever is provable from a contradiction.)

In passing it is worth noting that if we are using the Horn Clause subset and interpreting negation as failure, then we no longer have non-monotonic system. Consider the case where KB consists of a single clause $P \leftarrow$ not Q. From this KB alone P will be provable, since Q will fail. But if we extend KB by some Δ which contains Q, then P will no longer be provable. This is another emphasis of the need to make the closed world assumption if negation as failure is to be sound with respect to classical logic: making this assumption effectively rules out the possibility of extending the KB, and so the question of non-monotonicity is prevented from arising.

Despite this, workers in AI insist on the need for the systems to reason in a non-monotonic fashion. One example of everyday non-monotonic reasoning is well illustrated by the following story which John McCarthy tells. Suppose someone went to a maker of bird cages and ordered a bird cage for his bird. The builder built a fine cage and delivered it to his customer who refused to pay the bill on the grounds that the cage had a roof, which was quite unnecessary since his bird, Tweety, was a penguin, and could not fly. The question arises as to who was in the right here: the bird cage builder has assumed, without being told, that the bird intended for the cage could fly; the customer's contention is that he would have said if a roof had been needed. I think most people would agree that the customer is wrong: if you doubt this, consider the case where the builder had left off the roof, and where Tweety had been a more typical bird and flown to freedom.

Of course, on the basis of the information which had been supplied to him, the bird cage builder was not entitled to make any deductive conclusions about whether or not Tweety could fly. This being so, perhaps we might think he should have asked. But had Tweety been a normal bird, a canary, perhaps, this would have seemed a stupid question. This illustrates that not only do we use our knowledge of what is typically the case to draw conclusions to which we are not strictly entitled, but that we are expected to do so: a thing is normally only thought worth mentioning if it is not obvious, that is, if it is unusual or unexpected in some way.

This drawing of conclusions on the basis of insufficient evidence does not arise only in connection with what is normally the case. If we meet someone and they have a runny nose, and are sneezing, we may well come to the conclusion that they have a cold. We might ask them how long they have been suffering with the cold, only to be told that it is not a cold at all, but hay fever. That we were wrong and will need to retract our inference shows that we made the conclusion with insufficient evidence: strictly we should have eliminated the other explanations for the symptoms before making the conclusion. But this sort of thing is another everyday occurrence, and we will not be thought to have done anything out of the ordinary. In practice, deduction often requires far more information than we can expect to be supplied with.

There are thus two ways in which we may draw conclusions which are not deductively valid, and which may therefore later need to be withdrawn. The first relies on our "common-sense" knowledge of what is normally the case, and assumes that we will be told of any atypical features in a situation. The second is to make an inference to the best explanation: given a fact we have a natural urge to explain it, and will form the best explanation on the facts available, recognising that the explanation may need to be changed in the light of subsequent evidence. These two styles of reasoning are of independent interest to AI, and need to be distinguished as they may require different treatments. They are linked in that they both involve non-monotonicity, the retraction of conclusions, but the first is a kind of default reasoning, whereas the second may be termed abductive or plausible reasoning.

10.3.1. Non-monotonicity and frame representations

A slot-and-filler based representation with an inheritance hierarchy lends itself to an economical and powerful way of representing the information required to support default reasoning. Thus suppose we have a class bird which has some slot canFly. This slot can be filled with the value true in the class definition. When we form subclasses, such as finches and crows, this slot and its filler will be inherited by these classes, and further slots will be supplied with default fillers, so that the colour slot of crows will be filled with black. This process will continue down individual species such as budgerigars and ravens, where more default information may be added, say the disposition slot of raven with "malevolent". Now when we create an instance of one of these classes, say Tweety the raven, that instance will have many of its slots filled without any need for further information: we can answer that Tweety can fly, is black and has a malevolent disposition simply from being told that Tweety is a raven. Of course, the information given in this way may need to be withdrawn: we may in fact discover that Tweety is a benign albino, and need to record this information by replacing the default information in the disposition and colour slots by these values. Also this "cancellation", as the process of over-writing inherited information is called, may occur at the class level as well as the instance level: ostriches and penguins are birds, but will need to have the default value of false in the canFly slot.

This use of the representation is highly attractive, and indeed accounts for a good deal of the popularity of this sort of representation. It enables the builder of the representation to encode a lot of information as to what is normally the case in a very simple manner, and it provides a good model of the way human reasoning with regard to this seems to work, and it has the undoubted merit of allowing the system access to a useful information without the need to be explicitly given the information.

This ability to express default information using the inheritance mechanism seems to mean that such a representation has an expressive power beyond that of first-order predicate calculus. Within the latter notation we have good ways of expressing that all birds can fly, that some birds can fly, or that a particular bird can fly, but no way of expressing that birds normally fly, or birds typically fly, or most birds can fly, or that a typical bird can fly. Since a lot of what what we know, particularly about empirical domains, is in this latter form, it would seem that a representation which failed to provide a means of expressing such things would be expressively inadequate, and that this point is a very strong recommendation for the slot-and-filler representation. Its use does, however, create problems for interpretation of the meaning of the links in the inheritance hierarchy (see Woods [3] and Brachman [4]). The basic problem is that when we use a default value it is not clear whether we are saying that birds normally fly, a normal bird flies, a typical bird flies, most birds fly, or whatever. Since there are differences between these various things that we might be saying, this lack of clarity means that we cannot be certain as to what it is we are saying. Moreover, there is no guarantee that we will be applying a consistent interpretation for all the default values we use. Additionally, we need to be careful that we allow some of the default values to be uncancellable: if this is not so then we are unable to say that all members of a class have a particular value for an attribute, and this inability to express necessary characteristics would be a highly undesirable limitation on our expressive power. Of course, it is possible to take these points on board, and to implement and use the inheritance of default values in a disciplined way. In so doing, however, complications will be introduced and hard choices will need to be made so that the ease and simplicity which formed a large part of the attraction of the original idea will have been lost. The upshot is that some kind of formal model of the default reasoning will be required to render the representation sound, and such a model can equally be provided by an extended logic. Indeed there is a lot of current interest in developing such extended logics, and we will mention some of these in the section 10.33.

Further difficulties arise when we attempt to generalise the use of default values, because in many domains there is no natural default value for many of the slots, either because there are a very large number of exceptions, or because there is no "normal" value. As an example of the first, one could consider what default value one would assign to the colour slot of the class bird, and as an example of the second, consider the default value of the height of a person. In such cases we must either assign no default value at all, which will reduce the capability of the system, or we must be very careful that we are not prejudicing the interpretation of default values that we are trying to enforce.

10.3.2. Plausible reasoning

In summary, then, slot-and-filler representations and their use of default reasoning have drawn attention to an important aspect of human reasoning, and one which they seemed well adapted to implement. By contrast, inference to the best explanation is a style of reasoning more naturally associated with a rule-based representation. For there we have a set of rules (either production rules in a production system or clauses in a logic program), which represent sufficient conditions for the truth of some proposition. Thus we know that were the conditions true, they would force the conclusion to be true. So if we know that the conclusion is, in fact, true, the conditions can supply us with an explanation of why the conclusion is true, and so we have a reason to suppose that the conditions might be true. In this kind of representation, therefore, one of the rules or clauses will need to be selected as the favoured or normal explanation, which will be used unless there is reason to suppose that one of the conditions does not apply, in which case some alternative explanation must be found. This inference to the best explanation may be used in two ways: either to generate conclusions which will be assumed in default of a better explanation, or as a means of generating hypotheses which can then be tested. Used in the first method it becomes rather akin to the default reasoning from the normal case described above, whereas in the second mode it is rather a strategy which will be exploited by a problem-solving program.

Here, however, care must be taken about the basis for the rules that are used in this way. Examples where it is powerful are taken from areas where the relationships expressed in the rules are causal relationships. The explanation which is produced is consequently a causal one, and the plausibility derives from our underlying knowledge of cause and effect. But we may have rules which express other kinds of relationships, such as definitional relationships. We may therefore have a rule expressing the fact that a schooner has two masts: but to explain the two-mastedness of a particular vessel by assuming that it is a schooner would be a very odd way to reason. If abduction is used, therefore, care must be taken to apply it only where it is sensible to do so, namely to rules expressing causal relations.

There are further problems when we put information derived by default or abduction to use. Suppose we have Tweety the raven, and we start to make some inferences on the basis of the default values in the slots inherited from Tweety's class. This may leads us to draw a number of conclusions regarding Tweety and the way we should behave towards him. If, however, we find out that Tweety is exceptional in some way, a number of these conclusions will be invalidated. If, therefore, we are to retract some of our conclusions we need also a means of retracting any conclusions that we have drawn based on those conclusions. This is the problem of *truth maintenance*. Again this problem has attracted much attention of late. Since any disciplined method of truth maintenance will require a precise characterisation of the inferences allowed, we will first briefly review some proposed formal approaches to default reasoning.

10.3.3. Formal models of default reasoning

There are a considerable number of formal models of default reasoning (see Ginsberg [5]). All such models attempt to give a formal characterisation to statements of the form "normally P → Q". Such models will of necessity give rise to a non-monotonic logic, since we may become aware of abnormal circumstances blocking the default conclusion after we have already drawn it. Three different styles of formal model may be identified; the most famous of each will be given a brief mention here: more information may be found in the references given.

One approach is to extend the logical language and introduce a modal operator which means something like "it is reasonable to believe that". The best-known of these approaches is the autoepistemic logic of Moore [6]. He introduces an operator L, and has rules of the form "Lp ← q", which says that p is reasonably believed if q is true. The operator is made use of as follows. First produce the deductive closure of the axioms, A. Then if some statements Lp are in this deductive closure and their negations are not, add them. Do the same for statements of the form L − p. This procedure provides an extension of the original theory, A, which we will call T, and we may call the function which takes us from A to T th. Thus T = th(A). We may now repeatedly apply th to T, to give T' until T' = T, which will be the maximal set of rational beliefs on the basis of A. There may be more than one such maximal set: in such a case the intersection of the maximal sets is what should be concluded. Moore gives a semantics for this logic in terms of auto-epistemic models, and provides an alternative possible worlds semantics for his logic.

The second approach is to extend not the language but the rules of inference. This is best exemplified by the default logic of Reiter [7]. His additional inference rule requires expressions of the form A(X):B1(X), ..., BN(X) → C(X). This is meant to express that if A(X) is true and the negation of none of B1(X) to BN(X) is provable, then C(X) can be concluded. This differs from ordinary modus ponens in that the conditions in the antecedent divide into two, some of which must be true and the rest merely consistent to license the conclusion. Obviously, what is consistent with what is known can change as information arrives, and so the conclusions provable from such a rule may change also. Computability is achieved by regarding such default rules as a shorthand notation for the set of ground instances of the rule.

The third approach is McCarthy's circumscription [8], will represents an attempt to remain as close to standard first-order predicate logic as possible. Here default rules have the form p(x) & − ab(p) → q(x), where ab(x) states that x is "abnormal". Thus there are no new operators or inference rules, but rather a new class of predicate. Circumscription works by attempting to form a model which minimises the extension of ab, effectively allowing ab to be falsified by failure. Note that the ab predicate expresses abnormality with respect to the default rule in which it appears, and so each rule will require its own special ab predicate.

These treatments have attracted a good deal of analysis, and quite a lot is known about the different consequences of the treatments. What they have in common is the desire to represent information which is not deduced but assumed from background knowledge about what is typically the case within a formal system. This requires them to produce an extended logic to cope with defeasible conclusions. Note too that while these formal methods offer a logical account of defaults, they still require a specification of the circumstances under which these default conclusions should not be drawn. Thus we can only have real confidence in the conclusions if we have exhaustively represented the exceptional circumstances, so that they may not offer much in the way of practical gains over the less formal methods.

10.3.4. Truth maintenance

All of the above can only operate within a system which has a means of detecting and exploiting inconsistencies. Again we can point to three leading treatments of the topic of truth maintenance. Again this is not the place to give details: the interested reader should consult the references.

First we have the TMS of Jon Doyle [9]. This essentially requires the system to maintain a list of support for statements which can be inferred, (an *in-list*) and also a list of statements which support their negations (an *out-list*). A statement may be believed if statements on its in-list are believed and no statement on its out-list is believed. When a new statement is added, the in-lists and out-lists are considered and statements may be *inned*, that is included in the set of believed statements where they are supported by the new statement, or *outed*, that is removed from the set of believed statements, in consequence. This treatment has the consequence that the set of believed statements is always forced into a single consistent state. A further, undesirable, result is that a statement may out a number of currently believed statements which rest on its negation, although strictly the additional statement suggested that only one of these statements need be false. For example, if I currently believe a bird is a swan and a native of the Northern Hemisphere, discovery that it is black will force me to abandon one of these two beliefs. In Doyle's TMS, however, I must abandon both.

This last problem is addressed by McAllester's reason maintenance system [10]. This also identifies a set of statements which cannot be believed in the light of new information, but here the user is encouraged to select which member of this set he wishes to disbelieve in the light of the new information. The advantage is that more statements remain believed, but the price is that user intervention is required. McAllester's system does not require a special non-monotonic logic: normal logical deduction can proceed at the object level, and the truth maintenance takes place at a meta-level. Again the object is to maintain a single, consistent set of believed statements.

The currently most popular means of truth maintenance is the assumption based truth maintenance (ATMS) of de Kleer [11]. Here each inferred statement is recorded together with its justification. Each piece of information is asserted qualified by some set of assumptions termed an *environment*, and inferred statements are associated with the minimum environments which support them. The role of the ATMS is to calculate what can be believed in an environment, and to maintain a set of inconsistent (no-good) environments. This method does not regard statements as in or out, and so does not force a single consistent state: instead it keeps and reasons with all consistent environments, and relates statements to the environments which support them. A consequence of this is that far less inference has to be repeated, and switching environments is a quicker process than in the other two treatments. On the debit side, it does involve a good deal more housekeeping: Another advantage of this treatment is that it can be given a monotonic characterisation, since any conclusions are always conditional on an environment, and so it does not necessarily force adoption of a non-monotonic logic, since conclusions are drawn only conditionally on the basis of the assumptions current. Such conditional conclusions are purely deductive and will never need to be retracted, although those believed at a given time will depend on what is in the current set of assumptions. Thus the inference is deductive and monotonic, although the system considered as a whole may behave non-monotonically by changing environments.

10.3.5. Non-monotonicity and PROLOG

Before leaving this topic we can briefly look at how this bears on logic programming, using PROLOG. It will be recalled that the interpretation of negation as failure in such systems renders them non-monotonic. An alternative way of viewing matters is to consider that every predicate has a default value, in that it will default to false. It is also possible to write clauses that will make the default value for a predicate true by including a fact for a predicate with variables as its parameters as the final clause of a procedure for a predicate, but this requires that we be able to express the falsity of the predicate, otherwise it will be universally true. But if we adopt one of the approaches to negation discussed earlier, this can be done. Consider a PROLOG representation of the fact that most birds can fly:

    canFly(X,no):-penguin(X).
    canFly(X,no):-ostrich(X).
    ...
    canFly(X,yes):-bird(X).
    canFly(X,no).

Read literally, of course, the penultimate clause would state that all birds can fly, and the final clause that nothing could fly. However, given that we know that these clauses will be reached only after all the preceding clauses have been tried, the effect of the clause is to conclude that a bird can fly if it does not fall under one of the explicit exceptions, and then that anything which is not a bird cannot. Thus the effect is to get default reasoning; the price that is paid is to lose the strictly declarative reading of the clauses, since they need to take account of their context and the computation strategy of the interpreter. This has many similarities with the use of specificity in production-rule systems to cope with general rules and exceptions to them, and it has the same advantages and disadvantages that we found in Section 5.3.6.

10.4. Inexact reasoning and rule-based systems

In the discussion of non-monotonic reasoning, it was noted that first-order predicate calculus was unable to express facts such as "most birds can fly". In fact a good deal of the information we hold about the world relates not to absolute truths, or even facts about the majority of the members of a class, but about what is likely. Thus, in some domains, particularly games with an element of chance such as backgammon or poker, much of the knowledge of an expert will concern probabilities. At its crudest, it is helpful in poker to know that a straight is more probable than a flush in five-card games, whereas the reverse is true in seven-card games. The result of applying knowledge of this sort is not commitment to the truth of some proposition, but a degree of confidence in the truth of the proposition: thus if I hold a straight in five-card stud poker I am fairly sure that it will be the best hand, and much less sure if the game is seven-card stud.

In non-monotonic and default reasoning we have a way of recording what is the most likely case and reasoning as if it were actually true. Uncertainty is different in that we wish to be aware of all the possibilities and their relative likelihoods so that we can make informed decisions on the basis of this type of information.

Early expert systems focused quite strongly on this element of uncertainty in expert reasoning. In the case of MYCIN (see Section 9.5.1), where the field was medical diagnosis, the only real way to achieve complete certainty would be to let the disease run its course: but a doctor cannot do this, since it would then be too late to apply treatment. Instead, therefore, he must come to a judgement as to what is the most likely case and act on this, aware too of the strength of this likelihood and what the consequences of other possibilities are. In other words, he cannot simply ignore the other possibilities, as he would using default reasoning, but must balance them. Similarly, the domain of PROSPECTOR did not admit of certainty; only once the mine had been dug could the prediction be verified. But again deciding where to dig would require a judgement as to the likelihood of finding the mineral, so that an informed decision as to whether it was worth the gamble of digging could be made. Both these systems therefore needed to make uncertainty a feature of their reasoning, and the topic of representing uncertain information is an important one.

10.4.1. Sources of uncertainty

First, however, we need to recognise that uncertainty is not a homogeneous notion but that there are several possible sources of uncertainty. In the case of factual information we may firstly be less than certain because information is incomplete. This incompleteness may itself result from a number of things: it may be that the event has not occurred yet, as in the case when we feel sure that someone will not throw a double six in backgammon from our knowledge of the construction of the dice, but cannot be certain until the dice are cast, or it may be that we cannot for one reason or another perform the tests that are needed to provide confirmation, as when the field prospector believes the veins in a particular rock sample to bear a particular mineral but cannot perform the assay necessary to confirm this: in the worse case, we may not even be sure what tests we could perform to verify our suspicion. Secondly, our uncertainty may be because of some vagueness or ambiguity in the predicate involved. Thus while we may know a man of six feet six inches to be tall, we might be unsure whether a man of five feet eleven inches should have this predicate applied to him. Thirdly, there may be an element of irreducible uncertainty, as when we are trying to judge whether a person is sincere or not.

There is also uncertainty in the rules that we apply to this factual information in order to produce new conclusions. Experts, we are told, use heuristics, which are rules which are applicable in general but may fail in particular cases. This may be because there ought to be extra conditions for the rules to be applicable, of which our expert is unaware, or it may simply be an empirical association not adequately grounded in the causal chain. In such cases the expert may well have more confidence in some rules than in others: in the first case this will depend on the number and frequency of the omitted conditions. Alternatively these heuristics may fail because they refer only to the typical case.

10.4.2. Heuristics in backgammon

These different types of uncertainty may well be exhibited in a single domain. Suppose, for example, we were trying to represent the knowledge required to play an effective game of backgammon. (For those unfamiliar with the game, it may be helpful to say that a *blot* is a piece which can be sent back to the start if landed on ("hit") by one's opponent, and that offering a *double* means that the opponent must either refuse, immediately conceding defeat, or accept, in which case he must play on for twice the current stake.) This would involve the representation of many heuristics such as the following:

> BG1 Move blots to where they will not be hit
>
> BG2 If you must leave a blot you do not want hit within 7 of your opponent, go as close to your opponent as possible
>
> BG3 If you are a reasonable way ahead, but not very far ahead, offer a double
>
> BG4 If you are behind, but your opponent thinks that you are winning, offer a double

These are all examples of heuristics, things which are good things to do in general, but which may turn out badly in a particular game. BG1 is a counsel of perfection: very often it will be impossible to leave a blot where no possible dice throw would hit it. Following this rule, therefore, means that you must come to a judgement as to what your opponent's next throw will be, which you must necessarily make on the basis of incomplete information, and which you may choose to make on the basis of your knowledge of probabilities or on the basis of some kind of intuition. Moreover, this rule is defeasible in another sense: sometimes it is desirable to leave blots which will be hit, and the game situation may dictate that the rule be transgressed.

BG2 is firmly grounded in probability, and represents a way of trying to achieve BG1. This is simply because 7 is the most likely throw with two dice (1 in 6), and the probability of a specific number decreases as it diverges from 7, so that there is a 1 in 12 chance of throwing 4 with two dice and a 1 in 18 chance of throwing 3. Again, of course, following BG2 will not ensure that your blot is not hit, but will give you the best chance on the basis of probabilities.

BG3 is difficult to apply because of the vagueness of the predicates involved. The general idea is that you want to be far enough ahead so that your opponent will refuse the double, allowing you to win immediately without worrying about the vicissitudes of the dice throws. But, if you are far enough ahead both to insure against bad luck and to give the possibility of winning a double game (where your opponent loses by so big a margin that he must pay twice the stake anyway), you do not want your opponent to escape with the loss of only a single game. Within these general parameters, however, there is considerable scope for judgement, and you may well be unsure as to whether the rule applies to a given situation.

The final rule, BG4, is an example of intrinsic uncertainty. There can be no way of verifying that your opponent thinks you are winning: you have to rely on an assessment of his character, body language and feel for the game, and make your best guess on the basis of this information.

The need to distinguish between these various types of uncertainty is important because they are so different in nature that it is unlikely that a single representation will be suitable for all of them. Thus a probabilistic approach is likely to be good for determining dice throws, but less effective for resolving vagueness. With these distinctions in mind, then, let us look at the problems and some of the ways that have been suggested of resolving them.

10.4.3. Treatments of uncertainty

The first thing that a treatment of uncertainty in a rule-based representation must enable us to overcome is that we may be uncertain as to the applicability of the conditions in rules. Thus we may have a rule like

$$((P \mathbin{\&} Q) \mathbin{\&} (R \vee -S)) \rightarrow T$$

and be perfectly confident that it applies in all circumstances. Thus if we knew that P, Q and one of R and $-S$ were true, we could be certain of T also. But in practice we may be certain of P, but regard Q as possible, R as probable and S highly improbable. So how are we to combine this uncertainty to come to a conclusion as to the likelihood of T? The answer is that we must have some way of combining the various propositions when they appear conjunctively and when they appear disjunctively, and of the likelihood of not-S, given an assessment of the likelihood of S. In other words, given a complex proposition containing some uncertain elements, we must have a means of evaluating the certainty of that complex proposition.

The second thing a treatment of uncertainty must give us is a means of propagating uncertainty attaching to rules. Thus we may have a rule of the form $P \rightarrow Q$, and be perfectly certain as to P being true, but still less than certain about Q because we are not fully certain of the rule. And if we are less than certain of P as well, we need to have a means of combining these different uncertainties before we can have an opinion about Q.

Thirdly, a full treatment of uncertainty will allow us to combine the evidence of several rules. Thus if a number of heuristics point in the same direction, we may be readier to believe their conclusion than if a single heuristic pointed this way. So we do need a way of coming to an opinion about the conclusion taking our heuristics as a whole, rather than only individually.

We can therefore say that treatment of uncertainty can be characterised by the answers it gives to the following five questions, which can then also serve as a useful basis for the comparison of the various proposed treatments:

(i) How is confidence in evidence measured?
(ii) How is confidence in rules measured?
(iii) How is evidence combined?
(iv) Given uncertain evidence and an uncertain rule, what confidence can be placed in the conclusion?
(v) Given the same uncertain conclusion from several rules, what overall confidence should be placed in that conclusion?

10.4.4. Use of probabilities

A number of proposals for answering these questions have been made. One approach draws on standard probability theory. The idea here is that the likelihood of an event is expressed as a number between 0 and 1, 0 representing impossibility and 1 representing certainty. The conjunction of two events with probabilities p and q is p∗q, and the probability of an event of probability p not occurring is 1 − p. From this we can work out that the probability of one or both the events occurring is (1 − the probability of neither occurring), i.e. 1 − (1 − p)(1 − q), i.e. p + q − p∗q. Equally we can say that the probability of the exclusive disjunction of the two events is (the probability of the inclusive disjunction − the probability of both occurring together), i.e. p + q − 2∗p∗q. All of this shows that we have a good basis for answering the first and third of our questions. But the second and fourth of our questions pose more problems since standard probability theory does not really have a notion of rules, and so we cannot really assign probabilities to rules. We could, of course, using the kind of reasoning applied to disjunction, get a probability for a material implication P → Q, but that would defeat our object since it would require that we knew the probabilities of both P and Q. There is, however, an alternative means of propagating probabilities used in standard probability theory, namely Bayes' theorem.

Bayes' theorem uses the notion of conditional probability, namely the probability of an event occurring given that some other event has occurred. To give an example, we may wish to know the probability that a person is suffering from a certain disease given that he exhibits a certain symptom. This would be the conditional probability of the disease given the symptom, which we would write as P(d|s), where d stands for disease and s for symptom. Now Bayes' theorem tells us that P(d|s) is equal to P(d & s)/P(s).

The difficulty here is that information regarding P(d & s) may be very hard to come by. For this reason a more useful form of Bayes' theorem for our purposes is that

$$P(d \mid s) = P(d) * P(s \mid d)/P(s).$$

Here the necessity is that of providing an estimate of the probability of the symptom given the disease, and an expert may well be able to provide a reasonable estimate for this value. Implementing such a system does, however, require us to give a large number of such estimates since we will have to provide estimates of P(d), P(s) and P(s | d) for every disease and symptom that we want our system to be able to deal with. For a realistically sized diagnosis system, given the wide range of symptoms that people may exhibit for the various diseases, this may become intractable.

This problem is made worse when we consider how to handle the fifth question. This requires that we use a more general form of Bayes' theorem to allow us to take more than one symptom into account when making our diagnosis. This more general form is

$$P(d \mid s1,\ldots,sN) = P(d) * P(s1\&\ldots\&sN) \mid d)/P(s1\&\ldots\&sN).$$

Now this requires an estimate of the conditional probabilities of sets of symptoms given a disease, and of the prior probabilities of these symptoms —a number which grows exponentially. It might, however, be objected that we do not need an estimate of P(s1&...&sN), since this is equal to P(s1)*...*P(sN), by the reasoning given above. Unfortunately this is true only if the various symptoms are independent of one another, so that P(s1) = P(s1 | s2). In the sort of case we are dealing with this is unlikely, since the symptoms are all supposed to be related through some underlying model of the human body. Where we cannot assume independence, we must supply the necessary information.

There are then two types of problems with using probability to represent inexact reasoning. First, there is the sheer volume of probabilities that must be supplied, and second, there is the intrinsic difficulty in making these kinds of estimates. Using this sort of representation will produce very specific values for probabilities as output, but these will rely on the input probability estimates being accurate. In domains such as medical diagnosis, experts are unlikely to think in these terms and their estimates are likely to be subject to considerable error.

The best known application of these ideas is PROSPECTOR, which did indeed rely on estimates of conditional probabilities of various things given certain observable surface features being propagated in accordance with a version of Bayes' theorem. Even if the system succeeded, however, we should be rather sceptical that these numbers represented any kind of real probability, on which the validity of the inference relies, rather than some more vaguely defined intuitive notion of likelihood to which Bayes' style reasoning may be entirely inappropriate. There are, of course, domains which are tractable to the probability approach, particularly in the area of games involving cards and dice, but these tend to be rather artificial. The general applicability of the method is therefore dubious, as well as being difficult to implement because of the problems in assembling the basic information.

10.4.5. Certainty factors

A more usual approach is to adopt something like the certainty factors method used in MYCIN (Section 9.5.1). That system associated various propositions not with probabilities, but with a *measure of belief* (MB) and a *measure of disbelief* (MD), both being numbers between 0 and 1. These two measures did not necessarily sum to 1. MB and MD were then combined into a single number, called the *certainty factor* (CF), by subtracting MD from MB. Notice that this means that CF can range from -1 to 1. As evidence was accumulated, MB and MD would change, causing CF to increase or decrease according to whether the evidence supported the proposition or argued against it. Thus, abstracting from details peculiar to MYCIN, in answer to our first question, we can say that in this style of system that the confidence in the truth of a proposition is expressed by its certainty factor.

Next MYCIN also associated certainty factors with rules. Thus rules would be written as something like

> CFR1 IF Symptom 1 and Symptom 2
> THEN highly suggestive evidence (0.8) of Disease 1

Thus the confidence in rules was expressed by this number between 0 and 1 representing rule strength. 0 rather than -1 will be the lower bound since rules which are on balance disbelieved will simply not be included in the knowledge base.

Confidence in complex propositions was calculated in the following way. The conjunction of two propositions was given a CF equal to the minimum of the CFs of the component propositions, and the disjunction of two propositions a CF equal to the maximum of the individual CFs. The CF of the negation of a proposition would in MYCIN be simply $-1*CF$, i.e. MD $-$ MB. For example, suppose we have a proposition P with CF 0.6 and a proposition Q with CF 0.4. Now the CF of (P & Q) will be 0.4 and the CF of (P $\vee$ Q) 0.6, and the CF of $-$ P will be -0.6. This is a very straightforward way of combining the propositions, but is not obviously the best. The method of combining for conjunction assumes that in every case where the lower-rated proposition is true the higher-rated one will be also, and the method for disjunction that in no case where the higher-rated proposition is false will the lower-rated proposition be true. This seems optimistic with regard to conjunction and pessimistic with regard to disjunction.

The next question is how the CF of a set of conditions is combined with the rule strength to give a CF for the conclusion. Yet again a very straightforward method was used, namely to multiply the CF of the conditions by the rule strength. So if we consider CFR1 and assume that Symptom 1 has a CF of 0.3 and Symptom 2 a CF of 0.7, then the conditions as a whole will have a CF of 0.3, and the conclusion a CF of $0.3*0.8$, i.e 0.24. Again this seems to be recommended by its convenience rather than any other factor.

The final thing this treatment needs is a way of combining the evidence of several rules. Suppose that the knowledge base containing CFR1 also contains the following rule:

    CFR2 IF Symptom 1 or Symptom 3
                THEN weakly suggestive (0.3) evidence of Disease 1

Suppose Symptom 3 has a CF of 0.9. Now CRF2 will yield a CF of 0.27 for Disease 1. CRF1 it will be recalled gave a CF of 0.24 for this disease. The method use to combine the evidence provided by both rules taken together is to sum the independent evidence and subtract the product. So taken together the rules give $0.51 - 0.0648$, i.e 0.4452. Again this seems to be merely an implementational convenience, knowing that one must arrive at a result greater than either but less than the sum, rather than having any principled basis.

We can see, then, that this kind of reasoning allows us to extend a rule-based representation by answering our five questions listed in Section 10.4.3. The treatment is computationally very simple but intuitively rather unsatisfying, since it is hard to justify the methods used by reference to any human ratiocination. Another weakness is, of course, difficulties in getting experts to supply the rule strengths and the users to supply values for their certainty for the pieces of evidence used. However, one should consider the following interesting result of research [12] into the difference made by allowing varying degrees of precision in rule strength and CFs of evidence:

No of intervals	Same organisms and therapy	Different organisms	Different organisms and therapy
10	9	1	0
5	7	3	0
4	8	2	1
3	5	5	1
2	1	9	3

This shows that the performance of the system degrades significantly only when the number of different CFs drops below 3. This in turn suggests that great precision in CFs and rule strengths is unnecessary. This is good news as far as the practical construction of systems using this method is concerned, but reflects also the lack of a well-founded basis for the treatment. The whole approach seems to be very *ad hoc*, and reflects the need to work around a problem rather than a principled solution to the problem. Moreover, the difficulties in estimating the various CFs to be used are usually met by running sets of test data and adjusting the CFs in the light of the results achieved. This, of course, will work only if CFs behave in the remaining cases in a similar fashion to the tested cases, an unjustifiable assumption given the lack of a coherent theory. Thus the justification for such a treatment can only be any practical success it might achieve: and there is no a priori reason to think that it should succeed, other than that it seems to provide convenient approximations to what we feel must be the case.

10.4.6. Fuzzy logic

A third approach to inexact reasoning is based on the fuzzy set theory of Zadeh [13]. Zadeh's original work was a piece of mathematics designed to cope with the fact that the extension of some sets may be vaguely specified. Thus while "a positive integer less than 10" defines a set with a precise extension (the integers 1 to 9 inclusive), "a positive integer much less than 10", does not. 1 is clearly a member of this set and 9 is clearly not, but opinions may differ as the whether 2 to 8 are or are not members. Zadeh therefore proposed a "membership" function which would express the "grade of membership" of this set as a number between 0 and 1 inclusive. The idea is illustrated by Fig. 10.3. Each predicate will be associated with a particular

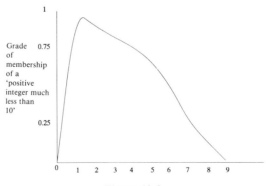

Figure 10.3

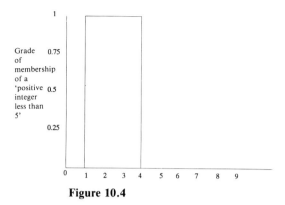

Figure 10.4

sort of curve for the range between the points at which it is definitely true and definitely false. In the case of a non-fuzzy set, this "truth profile", as the curve is called, will be a vertical line (Fig. 10.4). For vague predicates the precise shape of the truth profile may vary. The shape can be determined either by selecting from among a range of options, such as linear, step or parabolic, or can be derived by interpolating from some given values.

The idea has been extended by those interested in inexact reasoning to provide a general method of handling vague predicates. These too seem to have an extension which is imprecisely defined, and so fuzzy set theory would seem to provide a good way of handling them. Zadeh proposed means of handling the interaction between fuzzy propositions and the logical operators as follows:

For negation, member (complement of p) = 1 − member(p), where member (p) is the membership function applied to p (Fig. 10.5).

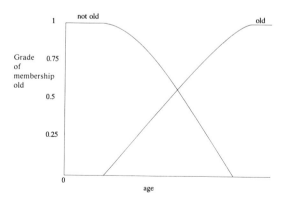

Figure 10.5 Treatment of negation in fuzzy logic. Not old = 1 − old.

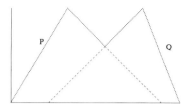

Figure 10.6 Treatment of disjunction and conjunction in fuzzy logic. ---, P&Q; ..., P∨Q.

For conjunction, member(p&q) = min(member(p), member(q)) (Fig. 10.6)

For disjunction, member(p ∨ q) = max(member(p), member (q)) (Fig. 10.6)

It will be noted that the treatment of conjunction and disjunction is the same as is found in rule-based treatments derived from MYCIN. We should notice that it relies on the independence of the propositions for its validity. An extreme demonstration of this need for independence can be seen by considering the application of this method to P& – P. Suppose P has a fuzzy truth value of 0.7, then – P will have a fuzzy truth value of 0.3 and P& – P will have a fuzzy truth value of 0.3 rather than the expected 0. Similarly with disjunction, where P ∨ – P will have a fuzzy truth value of 0.7 rather than the correct 1. Thus we need to assume independence of the constituent propositions, but this will be somewhat implausible in many cases.

As to the handling of uncertain rules, fuzzy set theory cannot help us. That was after all, only concerned to express grade of membership of a fuzzy set, not to provide a treatment of heuristics. Thus we must either take all our rules as certain, or adopt a method from some other area, like MYCIN's idea of multiplying the value of the conditions by the rule strength to give the value of the conclusions. The same, of course applies to our fifth question as to the handling of the merging of conclusions.

One interesting additional thing that fuzzy set theory gives us that the other treatments do not, however, is a means of handling "hedges". Given a vague predicate such as "young", we often qualify it by adverbs such as "very" or "quite", and these qualifying adverbs are called hedges. Fuzzy logic gives us a painless way of dealing with hedges by hypothesising that applying a hedge is to apply a further function to the result returned from the original membership function. The functions for the hedges need be defined only once, and then are applicable to all the predicates in our system. The results for "very young" and "quite young" can be shown diagrammatically as in Fig. 10.7. This means that given a truth profile for a predicate we can handle qualifiers in a uniform manner without adding to the knowledge base.

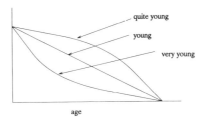

Figure 10.7 Hedges in fuzzy logic.

Zadeh's approach is based on a very interesting piece of mathematics, and has been used in some AI knowledge representations; (the commercially available REVEAL system [14], incorporates a means of handling fuzzy predicates). Again, however, it poses a number of problems (quite apart from the fact that to handle uncertainty in rules it must look to other approaches). These essentially concern the production of the truth profiles. How can we decide whether the truth profile for a given predicate is a straight line or a sine curve? What does it mean to say that it is 0.56 true that a person of five feet eleven inches is tall? And how could we decide whether it was better to say that a person of five feet eleven inches was tall was 0.58 true rather than 0.56 true? Fuzzy logic gives the appearance of precision, but does not give us a good grip on what the precision represents. By making the truth value capable of infinite precision, and forcing us to accept this by the continuous truth function, it ignores the fact that we cannot really decide matters to this degree of precision and so forces us to express things that we have no wish to express. On the other hand, it does make vagueness computationally tractable, and can give good practical results.

10.4.7. Qualitative representation of uncertainty

One final point about inexact reasoning: many people have objected to the need to attach numbers to the measurement of belief and disbelief, or probabilities, and have argued that uncertainty should rather be represented qualitatively, with such things as "possibly", "probably" and the like. In practical implementations, however, although these terms are used in the interface, when put to use by the inference engine they tend to be mapped on to numbers. Thus the appearance of qualitative reasoning is only an appearance: in practice it is just a treatment along one of the above lines, but with a restricted range of possible values. It may, of course, be of help to the user of the system, but it offers no theoretical gains.

10.4.8. Summary

In conclusion, one can say the following about inexact reasoning. First, there is a genuine need for such reasoning, and to represent the knowledge used in performing this sort of reasoning. Examples abound of reasoning with uncertain information to arrive at likely but uncertain conclusions. Second, formal techniques for capturing this style of reasoning have been developed in mathematics, most notably probability theory and fuzzy set theory. Third, these techniques are only applicable given the satisfaction of certain conditions that are in practice satisfied only in a limited number of domains. The consequence of this last point is that it is not obvious that we can apply these techniques in AI as providing a general treatment of inexact reasoning. The knowledge used by an expert will rarely take the form of precise probabilities once we move beyond stylised domains such as poker and backgammon, and so a representation of probabilities in, say the medical domain, is likely to distort the knowledge we are trying to represent. Thus we know that it is likely that an elderly person will need glasses, but this is not based on the fact that 95% of elderly people need glasses so that the probability of a given person falling into this group is 0.95, but on the fact that the eye is a "device" common to all people, and subject to a common ageing process which produces this kind of result. Thus when we apply a technique to an inappropriate domain we should be aware that we are doing so for pragmatic reasons and can justify the technique only by results and not by an appeal to the underlying theory.

This has led to the use of simpler more *ad hoc* approaches to inexact reasoning as found in the rule-based approaches derived from MYCIN. These approaches are not in conflict with the standard theories, but have as their main support their computational tractability. Again this will involve trying to assign numerical values to likelihood or certainty, which may be difficult to do, and hard to justify. Practically, the best approach seems to be to keep the number of categories available small, since little is gained by increasing the precision, and tuning them in response to results obtained from the system. Many of the implementations have achieved good results, but remain unsatisfying because of their lack of a sound theoretical justification. Finding such a justification is an open research issue.

10.5. Representation of control knowledge

Sometimes part of our knowledge about how to solve a problem consists of how we should apply our knowledge of the facts and relationships pertaining to the problem. Suppose we are given two lists, one giving the sex of a large number of people and the other giving a full list of child–parent pairs. Now suppose we want to find the son of a given individual, say Tom. Here it would clearly be sensible to use the second list to find Tom's children, and then to check whether they are male in the first list. The alternative would be to consider each of the males from the first list in turn and check whether they were the child of Tom. If we want to reflect this strategy in a PROLOG program we can do so only by the order of the body clauses:

> son(X,Y):-child-of(Y,X), male(X).

Because child-of appears before male in the body of the clause it will be tried first. But suppose now we want to test the truth of a query as to whether or not a given individual, say Hilary, is a son. In this case it would be sensible to look in the list of sexes first, since if she proves to be female we can stop immediately, whereas following the rule as it appears here we would have examined the second list unnecessarily. The point is that intelligent use of the rule depends on the nature of the problem, and this cannot be accommodated within the fixed execution strategy of PROLOG. We do represent implicitly the control knowledge by our order of clauses, but this order may not be optimal for every case.

We can say that in general, the most efficient strategy for using knowledge represented as a set of Horn Clauses is that if we have to choose between a number of clauses to solve a predicate we should choose the one most likely to succeed, since success will avoid any need to consult the other clauses, but when choosing a goal from the body of a clause we should choose the one most likely to fail, since a failure would mean that none of the remaining clauses in the body will need to be considered. The topic of query optimisation in PROLOG programs is well discussed by Warren [15].

Other considerations might lead us to want to use control knowledge in a program. Sometimes the correct execution of a clause will depend on whether or not a particular variable is instantiated when the clause is called. Consider the definition of the sum predicate in Section 8.4.6:

> sum(A,B,C):-integer(A), integer(B), C is A + B.
> sum(A,B,C):-integer(A), integer(C), B is C − A.
> sum(A,B,C):-integer(B), integer(C), A is C − B.

There we had to write three clauses, and use integer(A) to find which one was appropriate to use. If, however, it had been possible for the interpreter to be aware of the need to have the RHS of arithmetic assignments instantiated, and to exploit that awareness in its clause selection, the tests would not have been required, and a more declarative-looking program would have been possible, since this element of control knowledge would not have needed to intrude.

In Section 3.2.1 it was indicated that sometimes the correct problem strategy is to use data-driven search and sometimes it is to use goal-driven search. Without the ability to represent control knowledge, we have to opt for one or the other, but it may well be better to choose the direction of search once some aspects of the problem are known. This provides another motive for the explicit representation of such knowledge.

As a final example of where we might want to use control knowledge, consider the case where we have a repertoire of problem-solving techniques, the appropriateness of each dependent on some features of the problem. The selection of the technique to apply in a given case constitutes an important part of the knowledge of how to solve such problems.

There are then clear advantages to be gained from the representation and manipulation of control knowledge. Standard techniques include the ordering of clauses, although this is rather inflexible, and the inclusion of tests and flags in the condition parts of rules to indicate when they should be applied. The sum predicate above provides a small illustration, and the use of contexts in XCON, as described in Section 9.5.2, provides a more sustained example, of using conditions to represent control knowledge. These techniques do not, however, explicitly reason about control, rather they represent a convenient deviation from declarative representation. An alternative is to construct an interpreter which can use control information explicitly in order to produce a more flexible evaluation strategy. One of the more interesting attempts to explicitly represent and reason about control is provided by the MECHO system.

10.5.1. Representation of control knowledge in the MECHO system

The MECHO system, described in Ref. 16, is a program which is designed to solve problems in Newtonian mechanics, and is of interest here because of the way it guides its search by reasoning about the rules it may apply. This kind of reasoning is often called *meta-level* reasoning since it involves reasoning about reasoning. To enable this meta-level reasoning a special interpreter, MBASE, was written in PROLOG. This interpreter made use of the ordering of clauses as does PROLOG, particularly so as to implement a variety of default reasoning, much as described in Section 10.3.5. The interpreter, however, deviates importantly from the standard PROLOG interpreter in that it does not always employ PROLOG's chronological backtracking. In MECHO it is possible to control the depth of a call by specifying it to be one of three types: thus it may be a *database call*, which will succeed only if a corresponding ground literal is found in the database; an *inference call*, which will apply rules to solve the call in the usual way; or a *creative call*, which can generate place holders for intermediate unknowns, and so can carry on when an inference call would fail. Next there are a number of schemata which are used to generate contextual information, and to record the assumptions that can be made for a given problem. Finally, control is realised by the use of a number of *meta-predicates* which record the circumstances under which these various calls are to be used, and to determine such things as whether a particular equation can be used to solve for a given quantity. To give a feel for what meta-level predicates will be like, consider the following simplified example. We might have

   call__type(inf, Goal):-all__bound (Goal).

which would tell us to use an inference call to solve Goal if there were no unbound variables in goal. Or again, we might have a meta-predicate solve which would tell us how to deal with equations, partially defined like this:

   solve(U,Expression,Answer):-occurs-in(U,Expression,1),
         isolate__on__LHS(U,Expression,Answer).
   solve(U,Expression,Answer):-occurs-in(U,Expression,N),
         make-one-occurrence(U,Expression,Expression2),
         isolate__on__LHS(U,Expression2,Answer).

The first clause indicates that if the variable we wish to solve for occurs only once we can re-arrange the equation straight away, where if it occurs more than once, we will need first to collect these occurrences into a single occurrence. The actual meta-predicates in MECHO are rather more sophisticated, but these examples may give some of the flavour of what they do.

The underlying aim of the MECHO project was to produce a system for reasoning which would employ control predicates to determine how the rules in the system should be applied. As such it is an important piece of work and serves to highlight some of the limitations that come from using the rules according to a fixed and pre-determined strategy. The lesson that needs to be taken is that intelligence does involve knowing how to use knowledge as well as the knowledge itself, and that this requires the representation of meta-level knowledge, knowledge about knowledge. As with most of the topics in this chapter, the answers as to how this is best done are not yet available, and finding them is an active topic of research.

10.6. Time

The final issue of expressiveness that I shall consider relates to the treatment of temporal information. Predicate logic does not deal with time at all, its components being dereferenced so as to represent timelessly true (or false) propositions. Much of the information we want to use in an AI system, however, will require reference to time; the situation may be changing, either as a result of the actions of the system or exogenously, and we need to record information about how the situation at one time is related to the situation at other times. Such information thus represents a major challenge to the predicate logic paradigm.

If we consider production-rule systems first, however, we can see that they offer a relatively straightforward solution to these problems. The use of the working memory in production systems means that when a state changes the new state over-writes the old state, and working memory thus constantly represents the current state. Moreover, the rules governing these amendments of working memory provide a means of expressing how one situation follows from its predecessor. There are, however, some disadvantages: the chief one is that it makes the retracting of choices rather difficult, and so it is hard to backtrack to a previous choice point in a clean way. On the other hand, this mechanism does provide a natural way of representing a changing state; if the problem concerns a problem solver transforming an initial state into a desired one by affecting that state, one needs to reflect the changing status of propositions over time. The production rule

> IF (robot location (10,10) & (robot moves right) THEN (robot location (10,11)

is simple and effective as a means of expressing the effect of a particular movement in the position of the robot. But assuming that the robot is held to have only one location, it cannot express a truth in predicate logic, since whenever the antecedent is true the consequent will be false. Representing this in predicate logic requires that we be able to represent time, and this is not something to which predicate logic is well adapted. Some have sought answers in temporal logics and others have proposed first-order treatments (Kowalski and Sergot [17]), but it remains a difficulty for logic-based approaches. Frame systems share this feature with production systems, since new values of slots will typically over-write existing values. It is this feature that makes such systems so suitable for simulation applications, and less so for applications where backtracking is required.

10.7. Model-based representation

One of the most radical recent critiques of the whole range of rule-based knowledge representations has been put forward by proponents of systems based on what are called *deep models*. The term "deep" is used here to contrast with rule-based representations, which are said to be "shallow", This is an important new approach to knowledge representation which moves away from the idea of canonical paradigms for knowledge representation. It raises an enormous number of issues, and its long-term implications for AI may prove extensive. Here, however, I shall do no more than discuss the advantages that have been claimed for this approach.

To get a feel for what is meant here, consider the following simple rule that might be found in a system designed to diagnose faults in cars:

> IF the car does not start
> and the lights do not work
> THEN the alternator is faulty

This rule captures a degree of experience and expertise of a mechanic. His problem is that the car does not start, and he uses the functioning of the lights as a quick way of checking whether the battery is flat, and knows that a faulty alternator is a common cause of a flat battery. The rule thus encapsulates an understanding of the various causal relationsuips existing between the various components of the car, and depends on them for its validity, but the rule does not express these relationships, and does no more that record an empirical association.

Such a rule is probably not going to be good enough to use in the final system; the conditions will probably need to be refined. For example, if the lights had been left on overnight, then there is no reason to suspect a fault in the alternator as the cause of the flat battery, and so this might need to be reflected in the conditions. Additionally, there may be other reasons for suspecting a faulty alternator and this will need to be reflected by the addition of extra rules expressing further empirical associations.

As an application grows large and complex, however, this incremental refinement of the empirical associations may not be able to cope, both because the interactions between rules may become unpredictable, and because the knowledge will be increasingly difficult to elicit from the expert. Further, the empirical associations will be very task-specific: it is arguable that little knowledge of the domain has been represented, but rather a few heuristics that will assist a task in the domain.

An alternative is to produce a model-based representation of the domain. Instead of getting expertise from the expert, one could attempt to model the components and processes within the car engine. Such a model will express the causal relationships between the components, and so represent the underlying justification for the empirical associations of the rule-based representation, rather than the associations themselves. In theory at least, the model could be reasoned about "from first principles" and the empirical associations derived. Another key point is that such a model will separate structure and function: the model would represent that the rotation of the alternator produces an alternating current, but not that this meant that the battery became charged, although this would be derivable from the model. It is argued that to blur structure and function is to oversimplify and so miss out potentially significant features of the domain (such as the effects of the intervening wires in the example).

10.7.1. Advantages of model-based representations

The advantages claimed for model-based systems are firstly that knowledge is easier to acquire, since one can return to the original design documents of the device rather than cudgelling the brains of an expert. This may well be true of some domains, such as fault diagnosis, but is less clear in others. If we consider a model-based representation such as medical diagnosis, developing the model will be non-trivial since the system to be modelled is so complex and imperfectly understood, and may be far harder than going through an expert.

Secondly, it is claimed that such systems will be more robust. The rules obtained from an expert will be incomplete and coloured by his experience, so that the system may be inadequate in cases which are rarely encountered. The model-based representation will in contrast be neutral as to the frequency of situations.

Thirdly, it is argued that such systems will be more maintainable. Such a claim is a little ironic in that this was one of the advantages cited for rule-based representations over conventional computing programs. Despite this it has been found difficult to maintain such representations, particularly when the domain changes, invalidating some of the expert's existing empirical associations. Given a new model of car, for example, it is not always clear whether a given rule in the representation is still applicable, in need of modification, or no longer of value. The case should be easier with a model-based representation: clearly the model must be altered, but the areas in which it has changed should be clear given the correspondence between the model and the device modelled.

Finally, some other advantages have been mentioned, such as that model-based representation should be transferable to another system carrying out a different task on the same domain, and that explanations should be enhanced because they can refer to the underlying system rather than being a simple list of alleged empirical correspondences.

The approach is highly interesting but suffers from two major drawbacks. Firstly, it is only applicable to domains in which it is possible to construct a suitable model. This may well be possible in areas such as fault diagnosis on simple devices, but is very difficult in more complex domains. Secondly, there is the difficulty of reasoning with such models. Reasoning "from first principles", is not an easy matter, and computationally is very expensive. After all, even skilled people find it necessary to use the kind of associations used in rule-based representations in order to reduce the task to tractable proportions. Thus the kinds of techniques needed to support the use of model-based representations in practical systems remains very much a research area.

As an example of a model-based system one can cite Bratko's work on the identification of heart disorders from an electrocardiograph reading. This system modelled the heart as a mechanical device, and attempted to discover the associations between arhythmia and the ECG description. The program was successful, but far too inefficient to form the basis of a practical diagnosis system. Bratko therefore used his deep model to generate a set of shallow rules which could then be used as the basis of an ordinary rule-based system, which he reports performed acceptably well. Details of the system can be found in Ref. 18.

10.8. Conclusion

In this chapter we have seen that there are a number of outstanding issues which can present problems for the standard knowledge representation paradigms. We have also seen that the kind of issues that arise are common to the various paradigms, although they may offer different approaches to working around these problems. In many cases, differences in power between them represent things which they can be twisted to do, rather than which they do when considered in their purest forms. Any choice of representation depends as much on the nature of the domain, which makes the mapping of one paradigm more natural than the other, and the temperamental and stylistic preferences of the user of the representation as on any intrinsic differences in functionality. In this book the choices have been presented, and the issues raised: the reader must draw his own conclusions.

Exercises

10.1. True Horn Clauses do not permit negated clauses in their body, but logic programming systems such as PROLOG extend Horn Clauses to allow negated clauses in the body.
  (a) How do such systems interpret negation? Explain how such a treatment of negation might be implemented if PROLOG did not provide "not" as a built-in predicate, and the relation of this sort of negation to classical negation. Why cannot such systems implement classical negation?
  (b) Under what circumstances is this interpretation of negation sound? Give an example of where these circumstances might reasonably be presumed to obtain, and an example of where this presumption would not be reasonable, explaining the reasons for your choice in both cases.
  (c) Explain what is meant by a monotonic logic. Is PROLOG monotonic? Justify your answer.
10.2. Why do some people consider non-monotonicity an essential feature for AI systems? Are they right?
10.3. What is the standard form of explanation given by a rule-based system? What are the problems with it?

10.4. What is meant by abduction? What are its uses in the context of rule-based systems?
10.5. What problems with rule-based systems might be solved by the use of deep models of the domain?
10.6. How might we deal with incomplete or uncertain information in a rule-based system?

References

1. Ringland G.A., "Structured object representation—Schemata and frames", in *Approaches to Knowledge Representation* (ed. G.A. Ringland and D.A. Duce), Research Studies Press Ltd, Letchworth, 1988, p. 92.
2. Clark, K.L., "Negation as failure", in *Logic and Data Bases* (ed. H. Gallaire and J. Minker), Plenum Press, New York, 1978.
3. Woods, W.A., "What's in a link?: Foundations for semantic networks", in *Representation and Understanding in Cognitive Science* (ed. D. Bobrow and A. Collins), Academic Press, New York, 1975, pp. 35–82.
4. Brachman, R.J., "On the epistemological status of semantic networks", in *Associative Networks: Representation and Use of Knowledge by Computers* (ed. N.V. Findler), Academic Press, New York, 1979, pp. 3–50.
5. Ginsberg, M.L. (ed.) *Readings in Non-Monotonic Reasoning*, Morgan Kaufmann, 1987.
6. Moore, R., "Semantical considerations for non-monotonic logic", *Proceedings of the International Joint Conference on Artificial Intelligence*, (1983).
7. Reiter, R., "A logic for default reasoning", *Artificial Intelligence*, **13** (1980).
8. McCarthy, J., "Applications of circumscription to formalizing common sense knowledge", *Artificial Intelligence*, **13**, 27–39 (1980).
9. Doyle, J., "A truth maintenance system", *Artificial Intelligence*, **12**, 231–272 (1979).
10. McAllester, D.A., *An Outlook on Truth Maintenance*, MIT AI LAB AI-M-551, 1980.
11. de Kleer, J., "An assumption based truth maintenance system", *Artificial Intelligence*, **28** (1986).
12. Zadeh, L., "Fuzzy sets", *Information and Control*, **8** (1965); and "Fuzzy logic and approximate reasoning", *Sythese*, **30** (1975).
13. See Buchanan, B.G. and Shortliffe, E.H., *Rule Based Expert Systems*, Addison-Wesley, Reading, Mass., 1984, pp. 218-9.
14. REVEAL is marketed by ICL.
15. Warren, D.H.D., "Efficient processing of interactive relational database queries expressed in logic", *Proceedings of International Conference of Very Large Databases*, Cannes, 1981.
16. Bundy, A., Byrd, L., Luger, G., Mellish, C. and Palmer, M., "Solving mechanics problems using meta-level inference", in *Expert Systems in the Micro Electronic Age*, (ed. D. Michie), Edinburgh University Press, 1979.
17. Kowalski, R.A. and Sergot, M.J., "A logic based calculus of events", *New Generation Computing*, **4**(i) (1986).
18. Bratko, I., Mozetic, I. and Lavrac, N., "Automatic synthesis and compression of cardiological knowledge", in *Machine Intelligence 11*, Oxford University Press, 1989.

Bibliography

This does not attempt to be an exhaustive catalogue of relevant works. It contains all the items referred to in the body of the text, together with some others that the author has found helpful.

Aikins, J., "Prototypical knowledge for expert systems", *Artificial Intelligence*, **20** (1983).
Alty, J.L. and Coombes, M.J., *Expert Systems: Concepts and Examples*, NCC Publications, Manchester, 1984.
Bledsoe, A., "Non-resolution theorem proving", *Artificial Intelligence*, **9** (1977).
Bobrow, D.G. and Collins A., (eds.) *Representation and Understanding*, Academic Press, New York, 1975.
Boden, M., *Artificial Intelligence and Natural Man*, Harvester Press, Brighton, 1977.
Brachman, R.J., "On the epistemological status of semantic networks", in *Associative Networks: Representation and Use of Knowledge by Computers* (ed. N.V. Findler), Academic Press, New York, 1979, pp. 3-50.
Brachman, R.J., "I lied about the trees", *AI Magazine*, **6**(3), 60-93 (1986).
Brachman R.J., Fikes, R.E. and Levesque, H.J., "KRYPTON: integrating terminology and assertion", *Proceedings American Association for Artificial Intelligence-83*, Morgan Kaufmann, Los Altos, California, 1983.
Brachman R.J. and Levesque, H.J., *Readings in Knowledge Representation*, Morgan Kaufmann, Los Altos, California, 1985.
Bratko, I., Mozetic, I. and Lavrac, N., "Automatic synthesis and compression of cardiological knowledge", in *Machine Intelligence 11*, Oxford University Press, 1989.
Brownston, L., Farrell, R., Kant, E. and Martin, N., *Programming Expert Systems in OPS5*, Addison-Wesley, Reading, Mass., 1985.
Buchanan, B.G. and Shortliffe, E.H., *Rule Based Expert Systems*, Addison-Wesley, Reading, Mass., 1984.
Bundy, A., Byrd, L., Luger, G., Mellish, C. and Palmer, M., "Solving mechanics problems using meta-level inference", in *Expert Systems in the Micro Electronic Age* (ed. D. Michie), Edinburgh University Press, 1979.
Capper, P.N. and Susskind, R.E., *Latent Damage Law—The Expert System*, Butterworths, London, 1988.
Charniak, E. and McDermott, D., *Introduction to Artificial Intelligence*, Addison-Wesley, Reading, Mass., 1985.
Clark, K.L., *Negation as Failure, in Logic and Data Bases* (ed. H. Gallaire and J. Minker), Plenum Press, New York, 1978.

Clocksin, W.F. and Mellish, C.S., *Programming in Prolog*, Springer-Verlag, Berlin, 1981.
Crookes, D., *Introduction to Prolog Programming*, Prentice-Hall, Englewood Cliffs, N.J., 1987.
de Kleer, J., "An assumption based truth maintenance system", *Artificial Intelligence*, **28** (1986).
Doyle, J., "A truth maintenance system", *Artificial Intelligence*, **12**, 231–272 (1979).
Dreyfus, H., *What Computers Can't Do—A Critique of Artificial Intelligence*, Harper and Row, New York, 1972.
Duffin, P. (ed.) *KBS in Government 88*, Blenheim Online, Pinner, 1988.
Fiegenbaum E.A. and Feldman, J., *Computers and Thought*, McGraw-Hill, New York, 1963.
Findler, N.V. (ed.), *Associative Networks: Representation and Use of Knowledge by Computers*, Academic Press, New York, 1979.
Forgy, C.L., "Rete: A fast algorithm for the many pattern/many object pattern match problem", *Artificial Intelligence*, **19**, 17–37 (1982).
Gallaire, H. and Minker, J. (ed.), *Logic and Data Bases*, Plenum Press, New York, 1978.
Ginsberg, M.L. (ed.) *Readings in Non-Monotonic Reasoning*, Morgan Kaufmann, Los Altos, California, 1987.
Hammond P. and Sergot M.J., "A PROLOG shell for logic based expert systems", *Proceedings of Expert Systems 83*, pp. 95–104.
Hayes, P.J., "The second naive physics manifesto", in *Formal Theories of the Common Sense World* (ed. J. Hobbs and R.C. Moore), Ablex, New Jersey, 1985.
Hobbs, J. and Moore R.C., (ed.), *Formal Theories of the Common Sense World*, Ablex, New Jersey, 1985.
Hodges, W., *Logic*, Penguin, London.
Hughes, G.E. and Cresswell, M.J., *An Introduction to Modal Logic*, Methuen, London, 1968.
Hogger, C.J., *Introduction to Logic Programming*, Academic Press, London, 1984.
Jackson. P., *Introduction to Expert Systems*, Addison-Wesley, Reading, Mass., 1986.
Kowalski, R.A., *Logic for Problem Solving*, North-Holland, Amsterdam, 1977.
Kowalski, R.A. and Sergot, M.J., "A logic based calculus of events", *New Generation Computing*, **4**(i) (1986).
Lemon, E.J., *Beginning Logic*, Nelson, London, 1965.
Lovelace, A., "Notes upon L.F. Menabrea's sketch of the analytical engine of Charles Babbage", in *Charles Babbage and his Calculating Engines* (ed. P. Morrison and E. Morrison), Dover, New York, 1961.
McAllester, D.A., *An Outlook on Truth Maintenance*, MIT AI LAB AI-M-551, 1980.
McCarthy, J. and Hayes, P.J., "Some philosophical problems from the standpoint of artificial intelligence", in *Machine Intelligence 4* (ed. B. Meltzer and D. Mitchie), Edinburgh University Press, 1969.
McCarthy, J., "Applications of circumscription to formalizing common sense knowledge", *Artificial Intelligence*, **13**, 27–39 (1980).
McDermott, J., "R1: A rule based configurer of computer systems", *Artificial Intelligence*, **19**, 39–88 (1982).

Miller, P.L., *A Critiquing Approach to Expert Computer Advice: Attending*, Pitman Advanced Publishing Program, Boston, 1984.

Minsky, M., "A framework for representing knowledge", in *The Psychology of Computer Vision* (ed. P.H. Winston), McGraw-Hill, New York, 1975, pp. 211-277.

Moore, R., "Semantical considerations for non-monotonic logic", *Proceedings of International Joint Conference on Artificial Intelligence* (1983).

Nilsson, N.J., *Problem-Solving Methods in Artificial Intelligence*, McGraw-Hill, New York, 1971.

Nilsson N.J., *Principles of Artificial Intelligence*, Springer-Verlag, Heidelberg, 1982.

Pople, H.E. Jr., Myers, J.D. and Miller, R.A., "DIALOG: A model of diagnostic logic for internal medicine", *Proceedings of the 4th International Joint Conference on AI*.

Quillan, M.R., "Semantic memory", in *Semantic Information Processing* (ed. M. Minsky), MIT Press, Cambridge, Mass., 1968, pp. 216-70.

Reiter, R., "A logic for default reasoning", *Artificial Intelligence*, **13** (1980).

Rich, E., *Artificial Intelligence*, McGraw-Hill, Singapore, 1983.

Ringland G.A., "Structured object representation—Schemata and frames", in *Approaches to Knowledge Representation* (ed. G.A. Ringland and D.A. Duce), Research Studies Press Ltd, Letchworth, 1988, pp. 81-100.

Ringland, G.A. and Duce, D.A. (ed.) *Approaches to Knowledge Representation*, Research Studies Press Ltd, Letchworth, 1988.

Robinson, J.A., "A machine-oriented logic based on the principle of resolution", *Journal of the ACM*, **12**, 23-41 (1965).

Samuel, A.L., "Some studies in machine learning using the game of checkers", in *Computers and Thought* (ed. E.A. Fiegenbaum and J. Feldman), McGraw-Hill, New York, 1963.

Schank, R.C., "Conceptual dependancy: A theory of natural language understanding", *Cognitive Psychology*, **3**, 552-631 (1972).

Searle, J.R., "Minds brains and programs", *Behavioural and Brain Sciences*, **3**, 417-24 (1980).

Sebesta, R.W., *Concepts of Programming Languages*, Benjamin/Cummings Publishing Company, 1989.

Spirgel-Sinclair, S., "The DHSS retirement pension forecast and advice system", in *KBS in Government 88* (ed. P. Duffin), Blenheim Online, Pinner, 1988, pp. 89-106.

Susskind, R.E., "Government applications of expert systems in law", in *KBS in Government 88* (ed. P. Duffin), Blenheim Online, Pinner, 1988, pp. 185-204.

Turing, A., "Computing machinery and intelligence", in *Computers and Thought* (ed. E.A. Feigenbaum and J. Feldman), McGraw-Hill, New York, 1963.

Warren, D.H.D., "Efficient processing of interactive relational database queries expressed in logic", *Proceedings of International Conference of Very Large Databases*, Cannes, 1981.

Winston, P.H., "Learning structural descriptions from examples", in *Psychology of Computer Vision* (ed. P.H. Winston), McGraw-Hill, New York, 1975, pp. 157-209.

Woods, W.A. "What's in a link?: Foundations for semantic networks", in *Representation and Understanding: Studies in Cognitive Science* (ed. D. Bobrow and A. Collins), Academic Press, New York, 1975, pp. 35-82.

Wos, L., Overbeck, R., Lusk, E. and Boyle, J., *Automated Reasoning: Introduction and Applications*, Prentice-Hall, Englewood Cliffs, N.J., 1984.
Zadeh, L., "Fuzzy sets", *Information and Control*, **8** (1965).
Zadeh, L., "Fuzzy logic and approximate reasoning", *Sythese*, **30** (1975).

Index

Aduction, 23–4
ABox, 99
AI toolkits, 100
Aikins, J., 101
Allan Newell, 144
Alty, J.L., 169
Ambiguity, 16
Analytical Engine, 2
APES, 160
Aristotle, 28, 34
Arity of a relation, 37
Artificial intelligence (AI)
   beginnings of, 8
   definition of, 6–7
   motivations for, 7
   popular definitions, 4
   techniques used in, 7
   typical applications, 5–6
   use of term, 1–2
Assignments, 31–2
Assumption based truth maintenance (ATMS), 189
Attribute-value pairs, 162
Attributes, 95–6

Babbage, C., 2
Backgammon, heuristics in, 191–3
Backtrack, 42
Backward chaining, 70
Backwards reasoning, 45
BASIC, 18

Bayes' theorem, 194–5
Binary relations, 64
Bledsoe, A., 121
Blocks world, 45
Bottom-up resolution, 108
Bounded depth-first search, 46–51
Boyle, J., 121
Brachman, R.J., 85, 98, 99, 100, 101, 184, 210
Bratko, I., 209, 210
Breadth-first search, 46–51
Brownston, L., 77
Buchanan, B.G., 153, 169, 210
Bundy, A., 210
Byrd, L., 210

CADUCEUS, 156
Call, 135, 204–5
Canned text, 160
Capper, P.N., 167, 169
CENTAUR, 98
Certainty factor (CF), 196, 198–200
Charniak, E., 6, 7, 9, 140
Chess games, 11
Chess programs, 5–6
Choice point, 42
Clarity, 16
Clark, K.L., 175, 210
Class hierarchy, 91
Class-instance relation, 91
Class-subclass relation, 91
Classical logic, 39

Clausal form, 38, 105, 106, 173
Clauses, 38
Clocksin, W.F., 38, 124, 140
Closed world assumption, 175
Cognitive psychology, 6
Colby, K.M., 9
Completion of the database, 175
Computational tractability, 15, 19
Condition-action pairs, 63
Conditional probability, 194
Conflict resolution, 68, 71, 172
  algorithm for, 75
  strategies for, 72-4
Conflict set, 68
Conjunctive normal form (CNF), 38, 105, 106
Context, 154-5
Contradictions, 32
Control knowledge, representation of, 202-5
Coombes, M.J., 169
Counter-example, 31
Creative call, 204
Cresswell, M.J., 40
Crookes, D., 124, 140
CRYSTAL, 167
Cut, 132-4

Database call, 204
Data-driven search, 44-6
Declarativeness, 17-18
Decomposition of problems, 57-8
Deduction, 21-2
Deductive conclusion, 22
Deductive database, 127-9
Deep models, 206
Default information, 184
Default reasoning, 185
  formal models of, 186-8
Default values, 93-4
Defeasible rules, 74-5
de Kleer, J., 189, 210
Demons, 90
Depth-first search, 46-51
Destructive assignment, 18
Deviant logics, 40

Disjunctive normal form, 38, 105
Domain, 34
Doyle, J., 188, 210
Dreyfus, H., 1, 3, 8

8-puzzle, 52-9
Electronic brains, 3
ELIZA, 5, 9
EMYCIN, 99, 158
Entity-attribute-value triples, 63-4, 87, 98, 162
Epistemic adequacy, 14, 19
Equivalences between logical connectives, 33
Evaluation function, 53, 55-7
Existential quantifier, 35
Exotic logics, 39-40
Expert, use of term, 142-4
Expert system shells, 157-9
  escape to underlying system, 163-4
  typical facilities of, 160-4
Expert systems, 141-69
  advantages of, 166-7
  basic components, 148-52
  consultative, 145
  critiquing, 145
  domains of expertise, 145
  early, 152-7
  examples of new-style, 167-9
  heuristics in, 145-6
  incremental development, 151-2
  interactive, 145
  key features, 151-2
  possible features of, 146-7
  practical applications, 165-6
  trends in, 164-7
  use of term, 8, 142, 144-7
Expressiveness, 15-18

Fail, 132-4
Farrell, R., 77
15-puzzle, 52, 56, 57, 63
Fikes, R.E., 99, 101
Fillers, 87-8

Firing, 68
First-order predicate calculus, 20, 103, 120, 173
Flexibility, 85, 173
Forgy, C.L., 77
Forward chaining, 70
Forwards reasoning, 45
4-place relation, 37
Frames, 20, 86-100
  basic ideas, 87-9
  combined with rules, 98-9
  representations, 184-7
  use of, 89-90
Frege, G., 28
Fuzzy logic, 40, 198-201
Fuzzy set theory, 200

Game playing, 5-6
Games, 12
Ginsberg, M.L., 210
Goal-driven search, 44-6, 69
Goal state, 41

Haack, S., 40
Hammond, P., 169
Hayes, P.J., 13, 14, 25, 104, 120, 121
Heuristic adequacy, 14-15, 19
Heuristic notion, 59
Heuristic search, 51-5
Heuristics, 145-6
  in backgammon, 191-3
Hodges, W., 40
Hogger, C.J., 117, 121
Horn Clauses, 110-11, 125-7, 203
  expressiveness of, 173-82
  procedural semantics for, 116-17
Hughes, G.E., 40
Human mind, 6-7
Human problem solving, 57-60
Hume, D., 80

Identity, 37-8
IF . . . THEN format, 65, 68, 73, 75, 82, 89, 154-5, 162, 172

Incomplete information, 146
Incremental development, 76
Inductive reasoning, 22-3
Inexact reasoning, 190-1, 202
Inference call, 204
Inference engine, 148-9, 158, 163
Inference in semantic nets, 83-5
Inheritance hierarchies, 90-1
Inheritance mechanism, 184
Initial state, 41
Intelligence
  definition of, 3
  human versus machine, 3-4
Intelligent machines, popular views, 2-3
INTERNIST, 156-7
Intersection search, 85
Intrinsic uncertainty, 193
Intuitionistic logic, 40
Is-a class hierarchies, 91

Jackson, P., 38, 40, 79, 100, 169

Kant, E., 77
KEE, 100
Knowledge base, 21, 148, 152, 158, 161-3
Knowledge-based systems, 145
  use of term, 8
Knowledge elicitation bottleneck, 165
Knowledge engineer, 165
Knowledge representation, 171-210
  computational tractability, 15, 19
  criteria of adequacy, 13-15
  declarativeness, 17-18
  definition, 11
  epistemic adequacy, 14, 19
  examples of, 11
  expressiveness, 15-18
  heuristic adequacy, 14-15, 19
  introduction to, 11-25
  lack of ambiguity, 16
  major paradigms, 19-20

Knowledge representation *cont'd*
  metaphysical adequacy, 13-14
  need for clarity, 16
  notational convenience, 17
  object-orientated, 95-8
  purpose of, 12-13
  relevance, 17
  similarities between paradigms, 171-3
  uniformity, 16-17
Knowledgecraft, 100
Kowalski, R.A., 117, 121, 206, 210
KRYPTON, 99-100

Latent Damage Adviser, 167
Lavrac, N., 210
Lemon, E.J., 38, 40
Levesque, H.J., 99, 101
Linear input, 110
LISP, 84, 100, 137
Literals, 38
Locke, J., 80
Logic, 27-40, 103-21
  as representation paradigm, 120
  basics of, 27-8
  exotic, 103
  history, 28
  intuitionistic, 103
  non-monotonic, 103
  polyvalent, 103
  principles of, 21
Logic programming, 104, 111, 115, 125-7, 189
  as executable specifications, 117-19
  essential features of, 120
  foundations of, 105
  main ideas of, 117-20
  negation in, 174
Lovelace, A., 2, 9
Luger, G., 210
Lusk, E., 121

McAllester, D.A., 188, 210
McCarthy, J., 13, 14, 25, 182, 210

McDermott, D., 6, 7, 9, 140
McDermott, J., 169
Manipulation of representations, 20
Martin, N., 77
MBASE, 204
Measure of belief (MB), 196
Measure of disbelief (MD), 196
MECHO system, 204-5
Mellish, C.S., 39, 124, 140, 210
Meta-predicates, 205
Method of transformation, 38-9
Methods, 95-6
Miller, R.A., 169
Minsky, M., 86, 100, 176
Modal logics, 40
Model-based representation, 206-9
Models, 31-2
modus ponents, 32, 33
modus tolens, 32
Monotonicity, 33-4
Moore, R., 210
Most general unifier, 114-15
Mozetic, I., 210
Multiple inheritance, 94-5
MYCIN, 20, 153, 158-61, 163, 164, 190, 196, 200, 202
Myers, J.D., 169

Natural deduction, 32-3, 36
Negation
  as failure, 174, 182
    limitations of, 177-8
  in paradigms other than logic programming, 180-2
  treatments of, 178-80
Nielsson, N.J., 79, 100
Non-monotonic reasoning, 34, 182-90
Non-resolution theorem proving, 105
Normal form, 38
  varieties of, 105-6
Not, 134-5
Notation, 17, 34

Object-orientated knowledge representation, 95–8
Object-orientated programming languages, 95–8
Operators, 42
OPS5, 123, 153
Ordered search, 51–2
Overbeck, R., 121

Palmer, M., 210
PARRY, 5, 9
Part-of relation, 91
Pensions advice, 168–9
Plausible reasoning, 185–6
Polyvalent logics, 40
Pople, H.E. Jr., 169
Predicate calculus
  advantages of, 104
  basic of, 34–9
Predicates, 34
Predications, 34
Premises, 22
Prenex form, 112
Probability theory, 194–6
Problem-solving, 55, 97, 141–2, 186, 202
Procedural semantics, 116–17
Production memory, 67–8
Production rules, 20, 60, 63–77, 181, 185–6
  data-driven use of, 70
  form of, 63
  goal-driven search using, 69–70
  knowledge as set, 70
  variables in, 64–6
Production system
  components of, 66–8
  operation of, 68–75
  pros and cons of, 75–6
Programming languages, object-orientated, 95–8
PROLOG, 106, 110, 116, 119, 123–40, 173, 189–90, 203, 204
  arithmetic in, 136–7
  as AI programming language, 137–8
  as deductive database, 127–9
  features, 124–5
  features affecting control, 132–4
  features affecting database, 132
  features for input and output, 130–2
  for logic programming, 125–7
  implementation of negation as failure, 134–5
  non-logical features of, 129–37
  use of non-logical features, 135–6
Proof, 32
Propositional calculus
  basics of, 28–34
  notation of, 29–30
Propositional functions, 35–6
Propositional variables, 29
Prospector, 155–6, 159, 190, 195
Psychological justification, 85–6
PUFF, 99, 158

Quantifiers, 35–6
Query optimisation in, 203
Quillian, M.R., 80, 85, 86, 100

R1, 20
Recognise–act cycle, 68
reductio ad absurdum, 32–3
Referential transparency, 17
Refractoriness, 72
Refutation resolution, 108
Reiter, R., 210
Relational databases, 127–8
Relations, 36–7
Resolution, and predicate calculus, 111
Resolution methods, 105
Resolution principle, 105
Resolution strategies, 110
Resolvent, 106
Rete matching algorithm, 75
Retirement Pensions Forecast Advisor (RPFA), 168
REVEAL, 201
Reversibility of predicates, 119–20
Rich, E., 4, 6, 7, 9

Ringland, G.A., 210
RK611*, 154
Robinson, J.A., 105, 121
Roman numerals, 18-19
Rule-based systems, 190-1
Rule interpreter, 68, 172
Rules of inference, 32
Rules of natural deduction, 32
Rules of thumb, 59-60

Samuel, A.L., 169
Schank, R.C., 80, 100
Search for patterns, 58
Search limitations, 55
Search methods, 44-55
Search problems, 42
  examples, 42-4
Search space, 41-4, 55
Search techniques, 7, 41-61
Searle, J., 1, 2, 3, 8
Sebesta, R.W., 101
Semantic nets, 20, 79-86
  features of, 81-2
  inference in, 83-5
  origins, 80-1
Semantics, definition, 11
Sergot, M.J., 169, 206, 210
Set of support, 109
Shortliffe, E.H., 153, 169, 210
Skolem function, 112
Skolemisation, 111-13
Slot-and-filler representation, 184-6
Slots, 87-8, 94
Smalltalk, 95, 97, 100, 123
Specificity, 73, 85
Spirgel-Sinclair, S., 169
State, 41
State transitions, 42
Structured object paradigm, 98
Structured objects, 79-101
Subclasses, 91
Substitution instance, 113
Superclasses, 91, 94
Susskind, R.E., 167, 169
Syllogisms, 28, 34

Syntax, definition, 11

Tautology, 32
TBox, 99
Temporal information, 205-6
Thought experiment, 1
Time, 205-6
TMS, 188
Top-down resolution, 108, 110
Truth maintenance, 186, 188-9
Truth preserving, 22
Truth profiles, 201
Truth tables, 30-1
Turing, A., 4, 9
Turing test, 4-5

Unary relations, 64
Uncertain information, 146
Uncertain rules, 200
Uncertainty, 163
  qualitative representation of, 201
  sources of, 191
  treatments of, 193-4
Unification, 111, 113-14
Unifier, 113, 118
Uniformity, 16-17
Unit preference, 110
Universal quantifier, 35
User interface, 149-50, 160

Warren, D.H.D., 203, 210
Whatif?, 161
When-added procedure, 90
When-needed procedure, 90
Winston, P.H., 80, 100
Woods, W.A., 85, 98, 100, 184, 210
Working memory, 66-7, 71
Wos, L., 121

XCON, 76, 153-5, 164, 165, 204

Zadeh, L., 198, 210

The A.P.I.C. Series
General Editors: M. J. R. Shave and I. C. Wand

1. Some Commercial Autocodes. A Comparative Study*
   E. L. Willey, A. d'Agapeyeff, Marion Tribe, B. J. Gibbens and Michelle Clarke

2. A Primer of ALGOL 60 Programming*
   E. W. Dijkstra

3. Input Language for Automatic Programming*
   A. P. Yershov, G. I. Kozhukhin and U. Voloshin

4. Introduction to System Programming*
   Edited by Peter Wegner

5. ALGOL 60 Implementation. The Translation and Use of ALGOL 60 Programs on a Computer
   B. Randell and L. J. Russell

6. Dictionary for Computer Languages*
   Hans Breuer

7. The Alpha Automatic Programming System*
   Edited by A. P. Yershov

8. Structured Programming†
   O.-J. Dahl, E. W. Dijkstra and C. A. R. Hoare

9. Operating Systems Techniques
   Edited by C. A. R. Hoare and R. H. Perrott

10. ALGOL 60 Compilation and Assessment
    B. A. Wichmann

11. Definition of Programming Languages by Interpreting Automata*
    Alexander Ollongren

12. Principles of Program Design†
    M. A. Jackson

* Out of print.
† Now published in the Computer Science Classics Series.

13. Studies in Operating Systems
    R. M. McKeag and R. Wilson

14. Software Engineering
    R. J. Perrott

15. Computer Architecture: A Structured Approach
    R. W. Doran

16. Logic Programming
    Edited by K. L. Clark and S.-A. Tärnlund

17. Fortran Optimization*
    Michael Metcalf

18. Multi-microprocessor Systems
    Y. Paker

19. Introduction to the Graphical Kernel System (GKS)
    F. R. A. Hopgood, D. A. Duce, J. R. Gallop and D. C. Sutcliffe

20. Distributed Computing
    Edited by Fred B. Chambers, David A. Duce and Gillian P. Jones

21. Introduction to Logic Programming
    Christopher John Hogger

22. Lucid, the Dataflow Programming Language
    William W. Wadge and Edward A. Ashcroft

23. Foundations of Programming
    Jacques Arsac

24. Prolog for Programmers
    Feliks Kluźniak and Stanislaw Szpakowicz

25. Fortran Optimization, Revised Edition
    Michael Metcalf

26. PULSE: An Ada-based Distributed Operating System
    D. Keeffe, G. M. Tomlinson, I. C. Wand and A. J. Wellings

27. Program Evolution: Processes of Software Change
    M. Lehman and L. A. Belady

28. Introduction to the Graphical Kernel System (GKS): Second Edition Revised for the International Standard
    F. R. A. Hopgood, D. A. Duce, J. R. Gallop and D. C. Sutcliffe

29. The Complexity of Boolean Networks
    P. E. Dunne

30. Advanced Programming Methodologies
    Edited by Gianna Cioni and Andrzej Salwicki

31. Logic and Computer Science
    Edited by P. Odifreddi

32. Knowledge Representation. An Approach to Artificial Intelligence
    T. J. M. Bench-Capon

BRAULT